T0300904

The Instant Expert: Plastics, Processing and Properties

Dr Vannessa Goodship

WMG, University of Warwick

First published by Plastics Information Direct – a division
of Applied Market Information Ltd – in 2010
Republished by Elsevier

AMSTERDAM • BOSTON • HEIDELBERG • LONDON
NEW YORK • OXFORD • PARIS • SAN DIEGO
SAN FRANCISCO • SINGAPORE • SYDNEY • TOKYO

The Instant Expert: Plastics, Processing and Properties

First published by Plastics Information Direct - a division of Applied Market Information Ltd - in 2010
Republished by Elsevier

Elsevier
Radarweg 29, PO Box 211, 1000 AE Amsterdam, Netherlands
The Boulevard, Langford Lane, Kidlington, Oxford OX5 1GB, UK
50 Hampshire St, 5th Floor, Cambridge, MA 02139, USA

ISBN: 978-1-906479-05-3

Printed in the United States by Edwards Bros

Contents

Dedication

For Deb. Indubitably.

Preface

This book is made up of four distinct sections:

Part 1: Polymers and plastics

Part 2: Processing

Part 3: Properties

Part 4: The scope and applications of plastic materials

They all assume no prior knowledge of the subject. This book is not written to be used as an authoritative academic text, but simply as a step towards understanding the subject area. (Therefore, I apologise in advance for liberties taken with data to illustrate my points.)

The language of plastics can be baffling to anyone coming into contact with the business either occasionally or for the first time. This book aims to help the non-expert to understand their industry colleagues, customers and suppliers. However it should also serve as a useful reference book for anyone choosing or using plastics on a more regular basis. Finally, I hope it will provide a gateway to some of the fascinating and learned works produced on this subject. Suggestions for further reading can be found at the back of the book.

Acknowledgements

I would like to acknowledge the contributions of the following:

Deb and Nigal Goodship for taking the time to take hundreds of photographs for me (not all of which ended up in the final version).

Bethany Pearce for drawing some diagrams and for her suggestions.

Kylash Makenji for sharing his own research on decoration with me.

Erich Okoth Ogur for providing some processing diagrams for Part 2.

Okyndaye Erorewaen Bobby for some of the ideas on PLA.

Sally Humphreys for suggesting me for this work, Rebecca Dolbey for all her hard work editing it and Sandra Hall for typesetting it.

Vannessa Goodship

Part 1. Polymers and plastics

1.1 Introduction

Plastics – they are everywhere. You may love them, you may hate them but you cannot avoid them. As a thoroughly modern material plastic also seems to have been unfairly linked to all that is bad in our society. Our throwaway consumerism and unsustainable twenty first century lifestyles have left us with some serious problems and an awful lot to think about: our carbon footprint, global warming, saving tigers and bags for life. Waste, and plastic waste in particular, is a highly visible reminder of this. However, it is important to remember that plastic waste does not get there by itself (see **Figure 1.1**).

Plastics have transformed our society and still have the power to improve the quality of life for many people. Consider the plastics in your mobile phone, computer, television, and car; the medical implants improving lifespans, and the plastic syringes and tubes used every day in our hospitals.

A sustainable society is a phrase you hear often, with little thought for what it actually means. Do you want to sustain your lifestyle and environment? On the other hand, do you want to improve it?

For scientists, technologists, designers and engineers, there is enormous potential and scope for such advances in the field of polymers and plastics. The development of the next generation of polymeric materials, advances in processing technology and novel uses of a vast range of plastic materials could provide limitless opportunities to advance mankind.

This book will provide a glimpse into the sheer range and extent of a group of materials that like no other, impact on all of our lives.

Figure 1.1

Duck and discarded plastic bag
Source N. Goodship

1.2 Materials

Natural materials such as wood, rubber, cotton, starch, wool, leather and silk have long histories of use. All these materials also happen to be made of naturally occurring polymers. It is polymers that are also used to make the materials we know as plastics.

A polymer can be simply defined as a long chain molecule of repeating units as shown in **Figure 1.2.**

Proteins, enzymes, starches and cellulose are also natural polymer materials. Man-made synthetic polymers have a relatively short history in comparison and are made up by joining a series of small organic molecules together. These materials can be mass-produced relatively inexpensively and often perform better than their natural counterparts. Material properties can be tightly controlled and they have therefore replaced many of the older natural polymer materials in many applications.

Like metals and ceramics (e.g. materials such as glass, pottery, brick or cement), the properties of polymeric materials depend on the molecular make up of the material. The choice of one material over another for any specific application depends both on its use and the environment it must work within. For instance, a kettle must be able to withstand the heat from boiling the water, and plastic window frames must be able to withstand the seasonal changes of the weather. Due to the skills of design and engineering teams the choices involved in the material selection for new products need not enter the thoughts of the consumer. However, it becomes very apparent in the use of the wrong materials; *who would want a mobile phone made of concrete, a kettle made of paper, a football made of glass, toilet roll made of metal or an oven made of wood?*

One of the earliest examples of the use of commercial plastics suffered from a design and application problem. Cellulose nitrate was used to replace ivory billiard balls (ivory is another example of a naturally occurring polymer). Unfortunately, in this case the material had the unfortunate side effect of

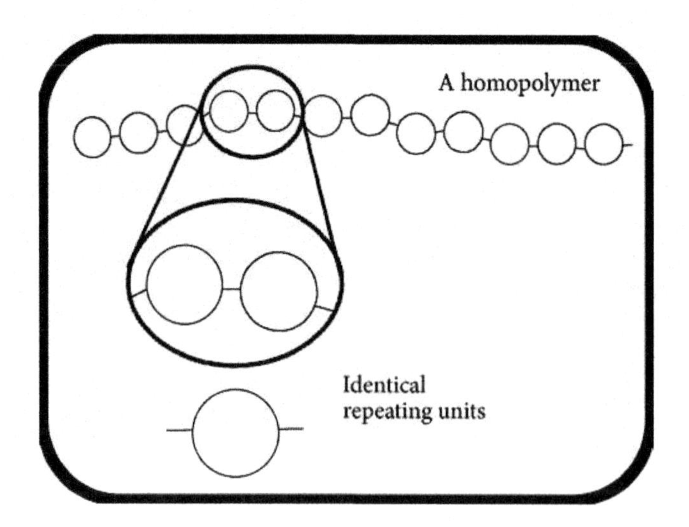

Figure 1.2

Simple polymer structure

occasionally exploding on impact due to the chemical mixture that was used[a]. Luckily, with developments in plastic materials and the skills of modern design engineers, this particular problem has long since been overcome.

The names of common material groups are familiar: wood, metal, ceramic and plastics. Within each of these material groups there are different types of each. Wood materials such as pine, oak, chestnut and mahogany. Metals such as steel, iron, copper, brass, gold, silver and aluminium. Likewise, there are also many different types of plastic. Some may be more familiar than others. You may have heard of polystyrene (PS), poly(vinyl chloride) which is more commonly known as PVC, polypropylene (PP) and polyethylene (PE). However there are many other types of plastics, and this variety allows designers to utilise plastics and replace the more traditional materials like wood, metal and ceramic in a whole range of applications. Traditional kettles were made of metal, traditional window frames were made of wood. In both cases and many others, plastics can be used to replace these materials. Plastic can also replace glass in certain applications, for example the lenses on our car lights are made of plastics. This glass replacement is often done using the plastic commonly known as Perspex.

Glass, wood and metal all have vastly different properties and uses, and this in itself attests to the sheer scope of plastics as alternatives to the more traditional materials. Sometimes however, the best choice of material is not so clear cut. Carbonated drinks are still sold in containers made from aluminium and glass, as well as plastic bottles. There are advantages and disadvantages to each of these materials in this case. These preferences are sometimes based on economic factors rather than just the materials themselves.

It can be seen in **Figure 1.3** that plastics have come to replace traditional materials in a whole range of applications and are used in diverse fields. They find application in packaging (e.g. bags, bottles, boxes), construction (e.g. cladding, building materials), automotive materials (e.g. engine components, bumpers, dashboard), electronic and electrical goods (e.g. bank cards, mobile phone covers, computer cases, TV housings) and medicine (e.g. artificial hips, syringes, medicine bottles). Many consumer products are made of plastics, such as toothbrushes, razors, toys and sporting goods. The reason for this success is the sheer range of materials available to meet designers' needs for virtually any application. To illustrate all the uses of plastics would be impossible (unless the intention was to produce a very big book). **Figure 1.4** illustrates just four applications for plastics, others will be highlighted throughout this book.

Hopefully it is already clear that plastics have become staple materials of modern living and are everywhere around us.

In 2006 the global production of plastics was 245 million tonnes, and about 25% of this was consumed within Europe (60 million tonnes). The United States, Mexico and Canada (NAFTA [b]) consumed 23.5%, China 14.5%, Japan 6%, Rest of Asia 17%, Latin America 4%, Middle East and Africa 7%. This global production total is larger than the production of metals worldwide, and another important factor when comparing plastics to metal is their much lighter weight.

a. A further application for this material also happened to be gunpowder.

b. These three countries are often known as NAFTA after the North American Free Trade Agreement – Members: America, Canada and Mexico

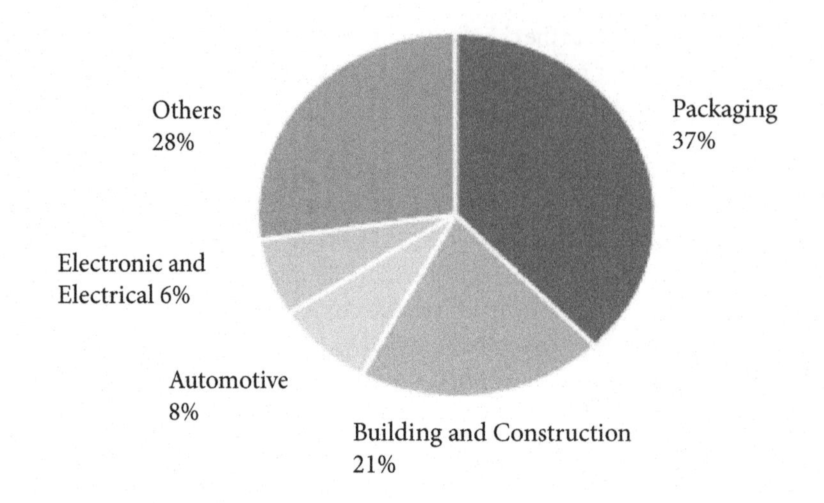

Figure 1.3

Plastic market sectors by weight

Reproduced with permission from PlasticsEurope 'The Compelling Facts about Plastics 2007'. www.plasticseurope.org

Figure 1.4

Plastic products: plant pots, toys, sporting goods, housewares

Source N. Goodship

Figure 1.3 shows where plastics are utilised; the packaging sector accounting for 37% of their usage, with building and construction as the second largest sector for the plastics industry.

In terms of sales volume, the plastics with the biggest share of the market are shown in **Table 1.1**. It can be seen that polypropylene is the most widely used polymer globally.

The materials used in the largest quantities are termed **commodity plastics** and this includes the plastic materials we are most familiar with everyday such as polypropylene (PP), the polyethylene family (HDPE, LDPE, LLDPE), polystyrene (PS), poly(vinyl chloride) (PVC) and poly(ethylene terephthalate) known as PET. The bulk of this commodity usage is in the packaging sector (**Figure 1.5**). As they are used in such high volumes, these materials also tend to be the cheapest materials.

Engineering plastics, which make up a smaller but still significant piece of the total market, are high performance materials able to provide enhanced properties (e.g. thermal, mechanical, chemical and electrical). They are utilised in markets such as construction (structural load bearing components), automotive (gears, bearings, cams, electrical switches) and industrial applications (chemically resistant materials, fittings and piping). The higher cost of these materials when compared to the cheaper commodity plastics is offset by the need for a performance product.

The class of engineering plastics includes plastics such as acrylonitrile-butadiene-styrene (ABS), polyacetal, also called polyoxymethylene (POM), polyamide, also known as nylon (PA), polycarbonate (PC), poly(phenylene oxide) (PPO), polysulfone (PSU), and the polyesters (PET and PBT). **Figure 1.6** shows some applications.

There are a further group of engineering thermoplastics that tend to be known as **Speciality Plastics** due to their specialised and low volume usage in specific harsh environments. These materials include fluorocarbon plastics such as PTFE, more commonly known as Teflon (**Figure 1.7**), poly(ether sulfone) (PES), polyarylate (PAr), poly(phenylene sulfide) (PPS), poly(ether imide) (PEI) and poly(ether ether ketone) (PEEK).

Some materials from the commodity group can also be considered as engineering polymers in certain circumstances. For example, if their properties have been enhanced by adding other materials and they can be utilised in engineering applications. PP can be modified with reinforcing additives to increase its strength and enable it to be used in this way. PET is also considered an engineering thermoplastic in its own right due to its high performance, however as it is used in such high volumes it is also considered as a commodity plastic.

Materials such as PVC and PS can also be used with engineering materials to produce high performance mixes called **blends.** Well known blends include ABS/PC, ABS/PVC and PPO/PS.

Table 1.1
Market share of common plastics

Material	Abbreviation	% of market share
Polypropylene	PP	19%
Low density polyethylene and linear low density polyethylene	LDPE and LLDPE (May also be seen as PE-LD and PE-LLD)	17%
Poly(vinyl chloride)	PVC	13%
High density polyethylene	HDPE (may also be seen as PE-HD)	12%
Polystyrene and expanded polystyrene (foamed)	PS and EPS	7%
Poly(ethylene terephthalate)	PET	7%
Polyurethane	PU	6%
Others		19%
Source: PlasticsEurope 'The Compelling Facts about Plastics 2007', www.plasticseurope.org		

Figure 1.5

Applications of commodity plastics (packaging and PVC guttering)
Source N. Goodship

Figure 1.6

Engineering plastics in action: ABS wing mirrors, PC helmet, PA wheel trims, PA hose connectors, ABS panel and PA buttons on a washing machine

Figure 1.7

Speciality plastics: PTFE tape

1.3 Polymers

So what exactly are plastics?

1.3.1 Definitions

A plastic must be capable of being made into articles by the act of moulding or forming. It contains a material called a polymer. It also contains other materials that are added to give the polymer increased functionality. Some of these materials will be introduced later in the section on additives but first a brief introduction to polymers is required.

Before we begin it should be noted that plastic and polymer materials are usually given a short abbreviation (e.g. PP, PS, PU see **Table 1.1**) as it can be seen that some of the polymer names are quiet long. Confusingly some materials are better known by their full name, and some by their abbreviation. However, it should be possible to cross-reference within the available text, tables or appendices if there is confusion. A further note should be made on the use of brackets in polymer names. Usually if the polymer consists of more than two words a bracket is used i.e. poly(ethylene terephthalate) and poly(vinyl chloride), however this usage is not always consistent within the literature or industry, especially if the material is well known. Therefore poly(vinyl chloride) and polyvinyl chloride are actually two ways of presenting the same material. The bracket convention will be used throughout this book.

The starting materials for synthetic polymers are usually materials such as oil, gas and coal. However plant-based materials such as cotton and vegetable oils can also be used. As the world moves towards finding alternatives to resources such as oil, it has been widely predicted in the future that more and more plastics will be produced from plant material. Like the question of biofuels this is an area of hot debate, much of it political, due to the instability of oil prices, habitat destruction and climate change concerns, and it is therefore beyond the scope of this book. However, the issue of plastic waste and biodegradability will be covered in later sections.

A simple definition of a polymer was given earlier. This idea will now be developed further:

A good definition of a polymer is a substance composed of very large molecules.

As these molecules are so big, they are also called macromolecules ('macro' is the Greek for large). For the word polymer we also need to thank the Greeks. A polymer is made up of a number of small distinct units called monomers ('monos' is from the Greek word for single). Once these are joined together they form polymers ('poli' is the Greek word for many) and the word 'meros' which means parts. As you will see 'many parts' is an apt description for polymeric materials.

There are many types of polymers, however one useful way to begin to categorise them is as:

- **Synthetic polymers**

Created by reacting monomers in controlled conditions, using chemical reactions such as polymerisation, polyaddition and polycondensation.

- or **modified natural substances.**

Created from materials such as cotton, celluloid, rayon, cellophane, and cellulose acetate.

This is illustrated in **Figure 1.8.**

You will note that the definition of a synthetic or natural polymer is somewhat confusing. They are all made from the resources of the planet. Polyurethane for example can be made from crude oil or vegetable oil. It is man-made and synthetic in both cases.

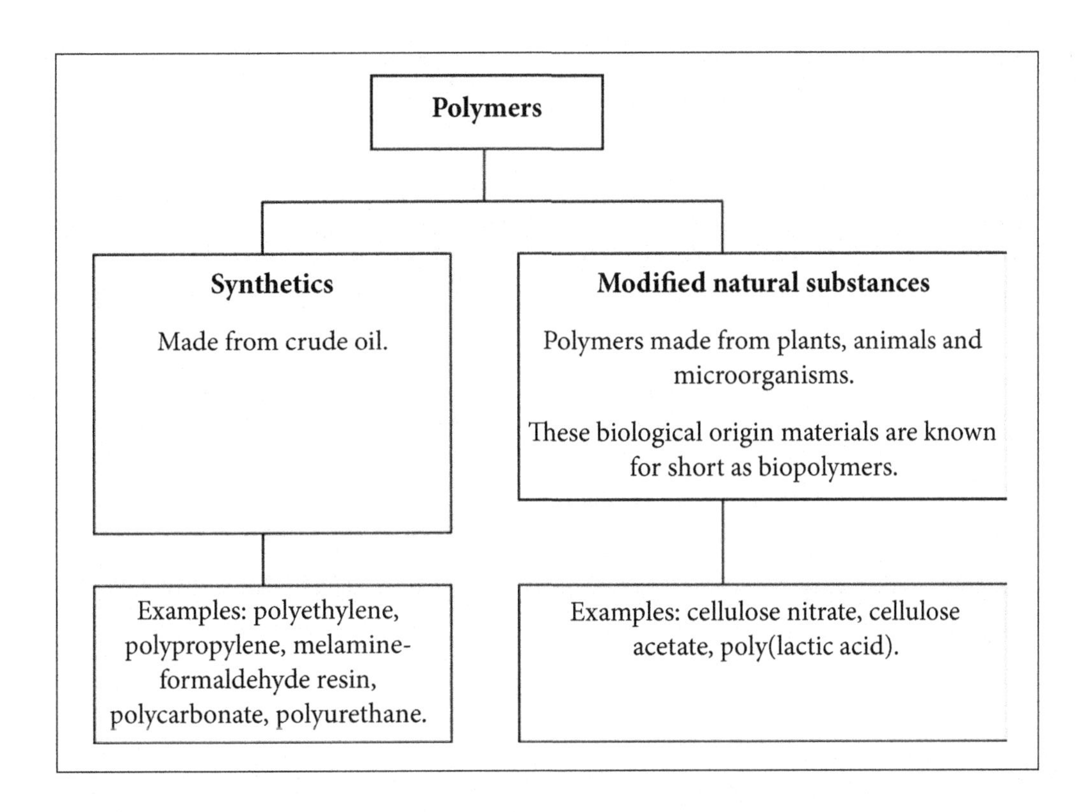

Figure 1.8
Types of polymers

1.3.2 The chemical representation of polymers

The monomers, which make up a polymer, can contain a number of different elements. One of the most common everyday polymers, polyethylene is made from the monomer ethylene (*for chemists, this is also known as ethene*). This is made up of two carbon atoms (C) and four hydrogen atoms (H). The joins between atoms in a monomer (or polymer) are described as bonds.

Just as water can be represented by the chemical formulae H_2O (the O in this case stands for oxygen), so ethylene can be shown as CH_2CH_2 or C_2H_4 (count the total number of Cs and then count the total number of Hs in the first example – both are the same.) During a process known as polymerisation, monomers are joined together to form polyethylene. The polymer chain is created as the carbon atoms join together to form a 'backbone' with the hydrogen atoms attached. One individual polymer chain can be made up of many thousands of monomer units, however a chain of useful polymer (in this context) may consist of 200-2000 of the monomers joined together. A chain of polyethylene is shown in **Figure 1.9** along with some different ways in which this molecule can be presented.

Often the polymerisation process produces a range of these chain sizes based around an average chain length. Both the average length of the chains (molecular weight) and the distribution of these lengths within any polymer material (molecular weight distribution) are factors in determining the properties of polymers made from the same monomer. A further factor in performance is the shape of the carbon backbone. Polymer chains can be linear as illustrated in **Figures 1.9** and **1.10**, or have branches, as is also shown in **Figure 1.10**.

The alignment of the chain and any branch structure affect the density of a polymer. As density is a measurement of mass per unit volume and branching reduces the chains' ability to pack together, branching therefore reduces the material density. There is less polymer (mass) in the volume of space (volume). Density is calculated by mass divided by volume. It is easy to see how this would be the

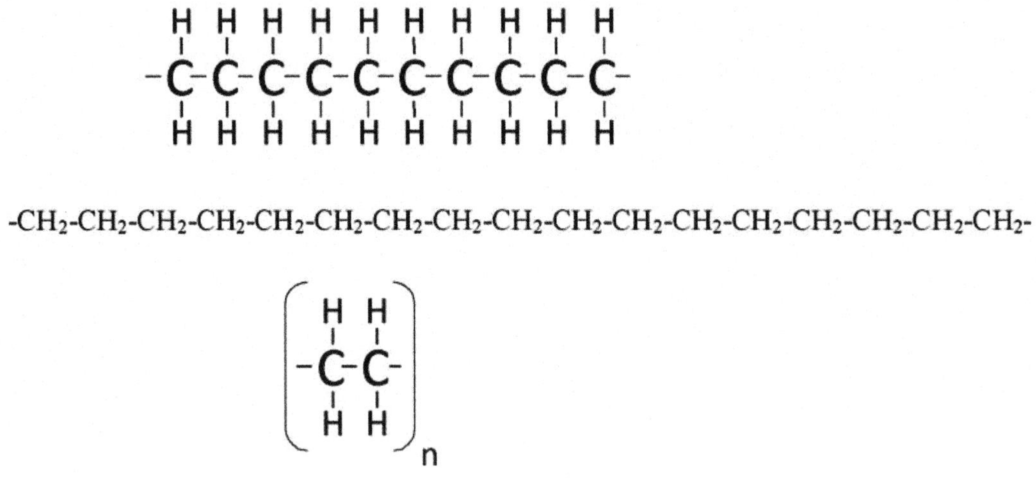

```
 H H H H H H H H H H
 | | | | | | | | | |
-C-C-C-C-C-C-C-C-C-C-
 | | | | | | | | | |
 H H H H H H H H H H
```

-CH2-CH2-CH2-CH2-CH2-CH2-CH2-CH2-CH2-CH2-CH2-CH2-CH2-CH2-CH2-CH2-CH2-

```
(  H H  )
(  | |  )
( -C-C- )
(  | |  )
(  H H  )n
```

Figure 1.9

Various methods of representing a linear polymer chain of polyethylene

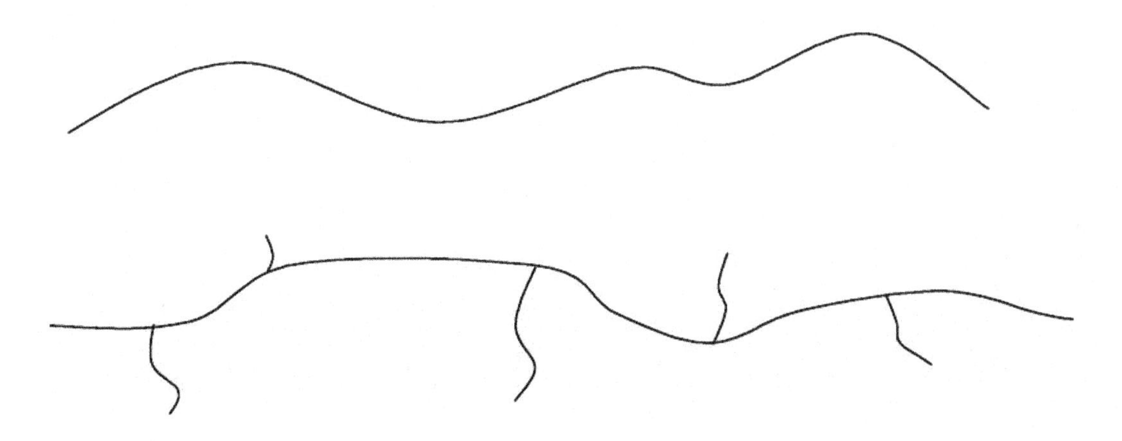

Figure 1.10

Simulated structure of linear (top) and branched (bottom) polyethylene

Table 1.2
Effect of branching on the properties of polyethylene

Polyethylene type	Melting point (°C)	Density (g/cm^3)
LDPE (branched)	115	0.92
HDPE (linear)	130	0.95

case as shown in **Figure 1.11**. Two well known types of polyethylene are high density polyethylene (HDPE) which has the linear structure and low density polyethylene (LDPE) which has the branched structure. The effect this has on density and is shown in **Table 1.2**.

The melting point is affected by how easily the polymers can align themselves. Again, it is easier to get chain alignment in the linear structure without the hindrance of the branches. Alignment leads to an increase in a property called polymer crystallinity, which raises the melting point. Crystalline and non-crystalline behaviour (called amorphous) will be returned to in more depth later. Variations in properties such as density and melting point will influence the applications a polymer may be used in, and also the processing conditions required to shape the polymer into a product.

In addition to having linear or branched structures, the polymer chains may also be crosslinked. In this structure the bonding produces a network of joins between adjacent chains. This is illustrated in **Figure 1.12**. The amount of crosslinking in this case has a marked effect on the properties of these polymers.

The type of bonding structure in the polymer has traditionally led to the distinction between two types of polymers namely thermoplastic (linear and branched) and thermoset (crosslinked). Whilst in the majority of cases this is true, there is always an exception to the rule! A polymer may become

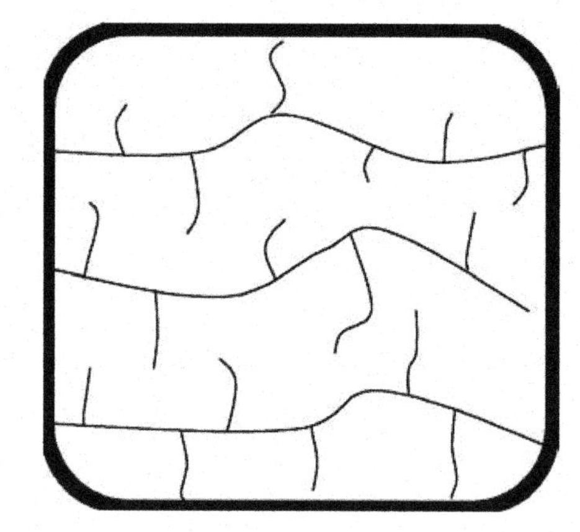

Figure 1.11

Effect of chain branching in reducing density

crosslinked without necessarily being a thermoset. One type of crosslinked thermoplastic called an ionomer is discussed later.

However, the general rule of thumb is that a thermoset is considered to be a crosslinked polymer and a thermoplastic polymer is one with a linear or branched structure.

As well as carbon and hydrogen atoms, polymers can also contain other elements such as nitrogen (N), chlorine (Cl), fluorine (F), or sulfur (S) atoms. The type of atoms and their arrangement is what differentiates the various plastics from each other. However they are all formed as long chains.

The longest polymer chains known today are found in DNA strands (yes, even our DNA is made up of polymers). Their monomer units are more complex than the materials that will be discussed here, but the chains can be the equivalent length of ten billion monomer units and would measure several metres long if uncoiled. The well known double helix structure of a strand of DNA is shown in **Figure 1.13**.

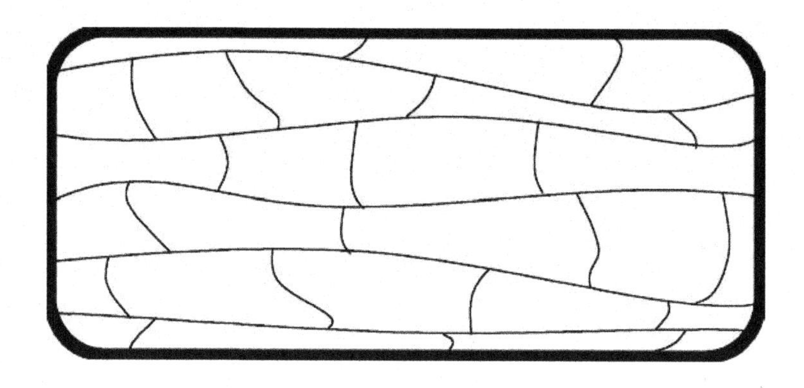

Figure 1.12

Crosslinked structure

Some common synthetic repeating units and the names of their resultant polymers are shown in **Figure 1.14**.

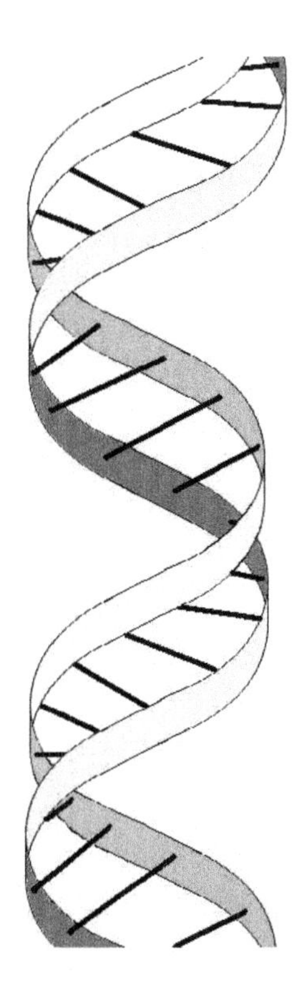

Figure 1.13

Illustration of a DNA strand

The polymers are created during the **polymerisation** process. There are two types of polymerisation reactions: **chain growth polymerisation** and **step growth polymerisation**.

Chain growth polymerisation proceeds as in polyethylene; a polymer chain grows as monomer units become joined to the end(s) of it, and there are no other products. This is because of the inherent chemical structure of the monomer; it is unsaturated[c] due to double bonds which exist between the carbon atoms.

At any time during the reaction the mixture will consist of just two species; monomer and the resultant growing polymer chains. These proportions varying as the reaction proceeds. (At the beginning it would be 100% monomer 0% polymer) and the reaction can be split into three phases: initiation, propagation and termination.

This type of reaction is used to create many of the commonest plastic materials such as polypropylene (PP), polyethylene (PE) as shown in **Figure 1.15**, poly(vinyl chloride) (PVC) and polystyrene

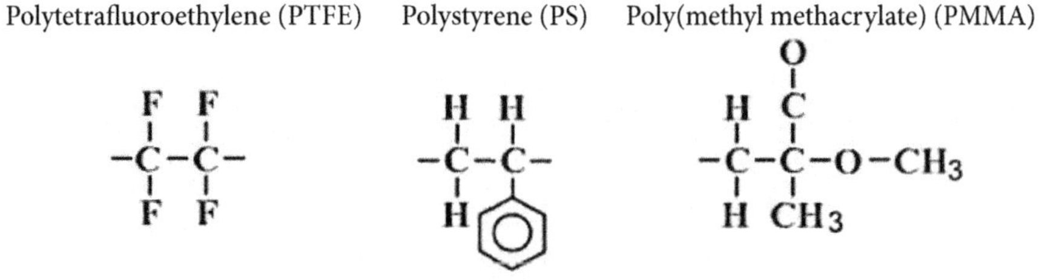

Figure 1.14

Common repeating units and polymer names

c. unsaturated – in this case the term applies to double and triple bonds between two carbons, that is where the carbon atom is not bonded to a maximum number of atoms and the capacity exists to attach further monomer groups. Carbons can each form up to four bonds per atom.

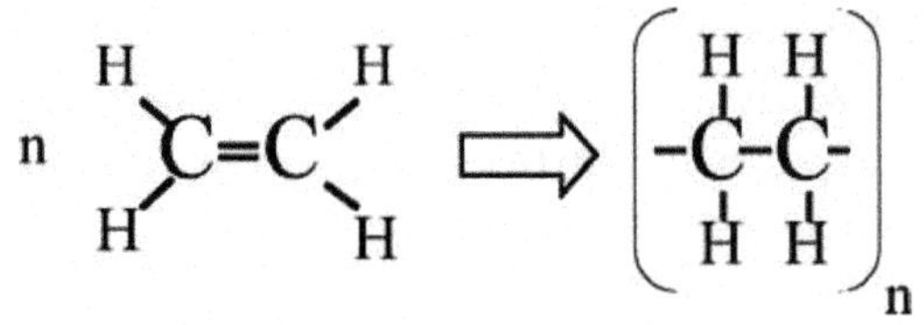

(PS). Chain growth polymerisation reactions can also be described as addition polymerisation.

$$n\ H_2C{=}CH_2 \Rightarrow \left(-CH_2-CH_2-\right)_n$$

Figure 1.15

Chain growth polymerisation of polyethylene (the n represents a variable number of monomer units)

Step growth reactions can proceed in two ways, one with by-products (polycondensation or condensation polymerisation) and one without (polyaddition or addition polymerisation).

In condensation polymerisation, the mechanism of chemical linking of the monomers involves reactive end groups on the monomers or growing chains that can bond together by expelling a small molecule, typically (but not always) water.

The principle of how this works is shown in **Figure 1.16**. An OH and an H group are each expelled from one chain to form H_20 as the by-product. Reactive groups remaining on the end of the chain then react with another monomer unit or chain to continue the reaction.

Typical polymers produced in this way include the nylons (polyamides) and the polyesters. A condensation polymerisation reaction is defined by two or more chemicals reacting to expel (condense) small molecules. This does not necessarily need to be water.

In step growth polyaddition, again there is a reaction of two functional compounds. However, in this case no side products are produced. Example polymers are polyurethanes and epoxy resins. Polymers made from both chain growth and step growth mechanisms are shown in **Table 1.3.**

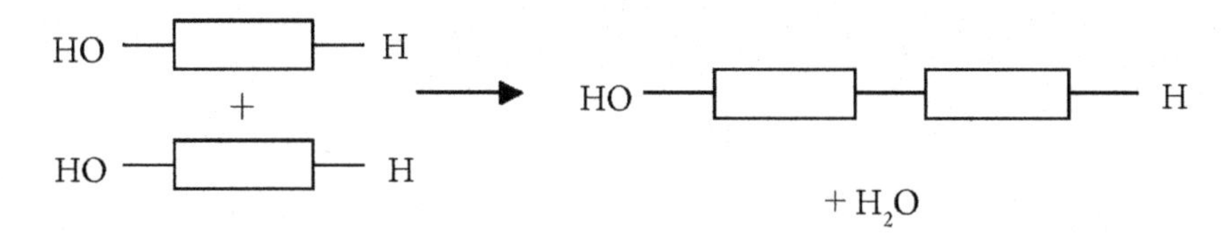

Figure 1.16

Principle of condensation polymerisation

Table 1.3
Polymerisation reactions and typical polymers produced

Chain Growth	Step Growth
Polyethylene (PE)	Nylon 66 or polyhexamethylene adipamide (PA 66)
Polypropylene (PP)	Polyethylene terephthalate (PET)
Polystyrene (PS)	Polyurethane (PU)
Poly(vinyl chloride) (PVC)	Epoxy resin

1.3.3 Isotactic, syndiotactic and atactic polymers

It has been shown that the arrangement of the branches on the polymer chain can have a pronounced effect on polymer properties. The placement of individual groups on the carbon backbone also exerts an effect. A further examination of this requires a consideration of a term called the **tacticity.** For this, it is necessary to return to the molecular structure of the polymer chain.

Ethylene can be represented as shown in **Figure 1.9.** However in many polymers there will be different atoms or groups of atoms bonded to the backbone in the place of some of the hydrogens. Examples could be a CH_3 group or another atom such as chlorine (Cl). The regularity and symmetry of this arrangement can significantly affect the properties. For the following examples the letter R will be used to represent the side group or atom that is not a hydrogen.

The tacticity of a polymer is defined by these relative configurations along the chain. They are termed stereoisomeric configurations. A summary of terms is given in **Table 1.4.**

If all the R groups are situated on the same side of the chain, the polymer is said to have an isotactic configuration (**Figure 1.17**).

If the R groups alternate from side to side (**Figure 1.18**), then it is said to have a syndiotactic configuration.

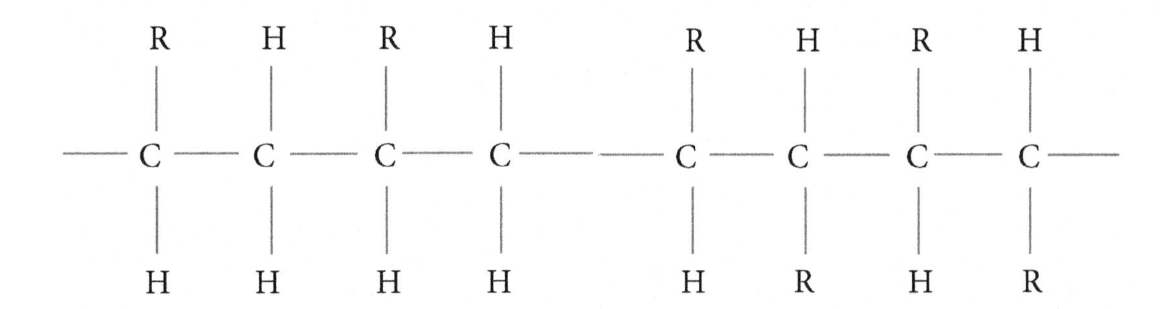

Figure 1.17

Isotactic configuration

Figure 1.18

Syndiotactic configuration

Table 1.4
Molecular configurations of polymer chains

Name	Position of R groups	Crystalline behaviour
Isotactic	R groups on same side of chain	Semi-crystalline
Syndiotactic	R groups alternate to either side	Semi-crystalline
Atactic	Random	Unable to crystallise

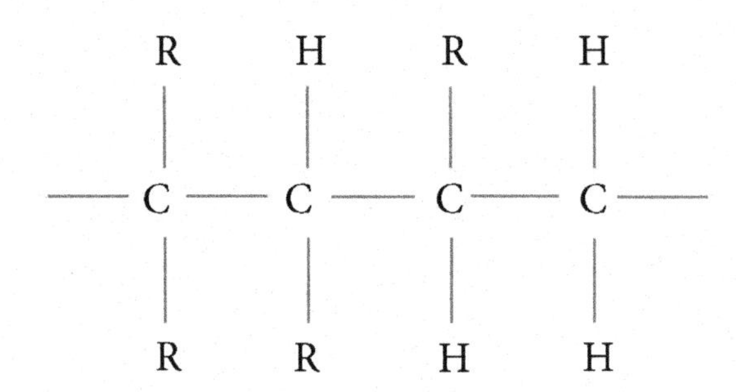

Figure 1.19

Atactic configuration

If the positioning is random, the term atactic is used (**Figure 1.19**). A polymer may not exhibit just one of these configurations, however a predominant form can be controlled during synthesis.

For example, polypropylene is commercially sold in all three configurations. However, the isotactic material is the one most commonly used.

Examples of atactic materials include PS, PVC and poly(methyl methacrylate) (PMMA), which is more commonly known as Perspex. Atactic materials are unable to crystallise.

An ability to crystallise also gives a distinction between two types of material groups, with those that can called semi-crystalline, and those that cannot called amorphous materials. One much used property of amorphous materials is their transparency and the term 'glass' is actually quite apt in this case. Semi-crystalline and amorphous materials will be discussed further in section 1.3.6.

1.3.4 Polymer chain coiling

Next it is necessary to consider what polymer chains actually look like. Imagine holding a very long piece of polymer chain consisting of the carbon backbone and hydrogen atoms (like a polyethylene chain). Would it be straight and rigid or would it be misshapen and flexible?

A long piece of cooked spaghetti is a good way to picture this.

In the carbon-carbon bonds, there can be small rotations around each of the carbon axes. However because there are so many of these carbon-carbon bonds, by the time they are all considered together a complex system of movements has occurred causing polymer coiling (like a spinal movement). This gives polymers flexibility. However this flexibility depends on the nature of the bonds that exist, and some bonds are less flexible than others, for example a double carbon bond is far more rigid than a single carbon bond. The atomic make up of the chain also affects the rigidity. Again, different combinations of atoms may have more or less flexibility than a carbon-carbon bond. It has been shown that polymers are made up of different atoms in linear and branched forms. This also affects a chain's ability to rotate and move. All these variations help to explain the different properties that polymers of the same family can have (polyethylenes for example) or different families of polymers, and hence the different characteristics of the plastics that are made from these polymers.

Given the coiled nature of polymer chains, the longer the length of individual chains the more entangled with other polymer chains they can be become.

Imagine now a whole bowl of long entangled spaghetti! (Spaghetti is a good representation of a bowl of linear polymer chains.)

The average length of the chains (molecular weight) in the polymer obviously has an effect on how entangled they become. Also remember the effect of how closely the chains are able to pack together with each other, which was highlighted in **Table 1.2** and **Figure 1.11**. Consider again the polyethylene family. Linear polymer chains can pack more closely than chains with a chunky branched structure. This affects the density of individual polymer structures. These ideas will be further explored in the discussion of amorphous and crystalline materials in section 1.3.6.

1.3.5 Copolymers

The polymers discussed so far have been produced by using either one monomer repeating unit, or several chemicals being reacted to produce a desired polymeric chain (as described in polycondensation reactions). These are the least complex ways that polymers can be combined and are called **homopolymers**.

However, more complex interactions can be used, for example polymers can be produced from two or more different monomers. These materials are termed **copolymers** and they are produced in order to generate materials with slightly different properties than is possible with a homopolymer. This is illustrated in **Figure 1.20**.

There are a number of types of copolymer structures. These are illustrated in **Figure 1.21**. Some materials such as PP can be bought in a range of versions. This is done to alter the properties of the polypropylene from those that can be produced from one monomer alone.

For example, the impact properties of a polypropylene homopolymer can be greatly enhanced with the inclusion of polyethylene in both the block and random configurations. The resultant properties however are very different, and these general property differences can be seen from the examples given in **Table 1.5**.

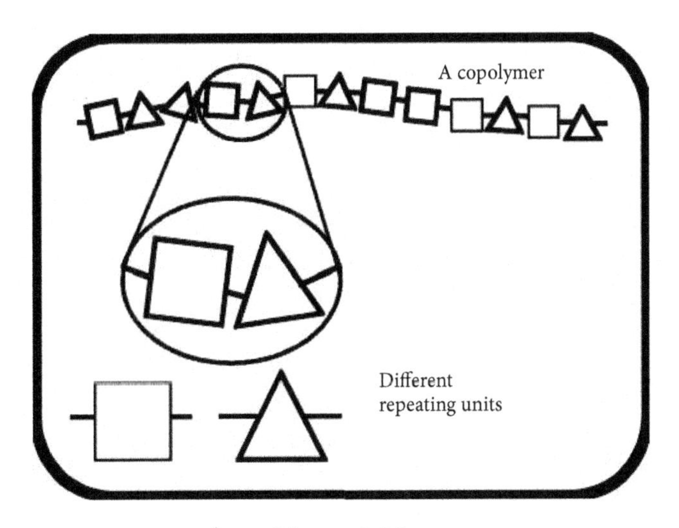

Figure 1.20

A copolymer with two repeating units

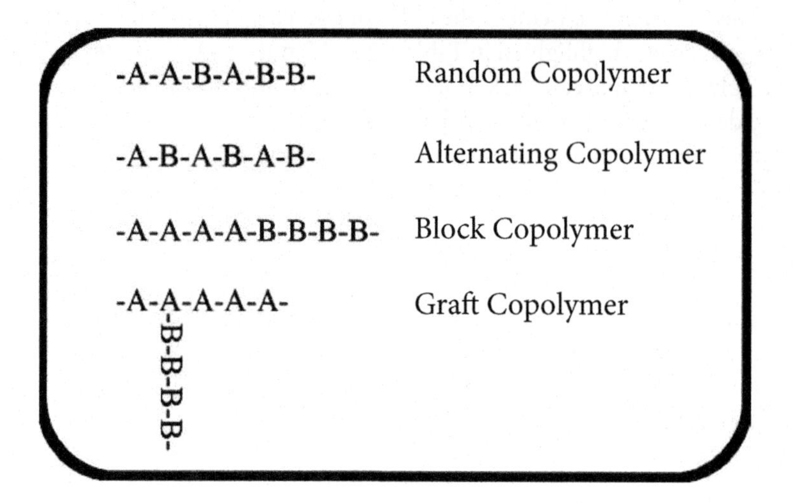

Figure 1.21

Types of copolymer structure (A and B represent different monomer groups)

Table 1.5
Some examples of common copolymers with different types of structures

Type of copolymer	Abbreviation	Full Name	Properties
Alternating	SMA	Styrene/maleic anhydride alternating copolymer	High heat resistance, low coefficient of linear thermal expansion (CLTE), good solvent resistance
Block	SEBS (This is a thermoplastic elastomer: see section 1.4.3)	Styrene/ethylene-butylene/styrene block copolymer	Hard degradation resistant elastomer, used for soles, tubing, baby teats, syringe seals
Graft	Hivalloy Reactor grafted amorphous + semi-crystalline materials	High value alloy (e.g. PP/PS, PP/PMMA)	Improved stiffness/ impact balance, low density, good chemical resistance
Random	PP copolymer (Ethylene is the most commonly used comonomer)	Polypropylene copolymer	High impact properties, higher stiffness, better clarity than homopolymer PP

Both materials on their own (polypropylene, polyethylene) when polymerised produce semi-crystalline materials with distinct melting points. A block copolymer of 50% ethylene and 50% propylene also has a clear melting temperature. However, synthesis of the same copolymer in random configuration results in the disappearance of the melting point giving 'rubber' behaviour. The resultant material is an elastomeric material called EPDM rubber. Elastomers are discussed in more detail in section 1.4.3.

Another common combination of monomers is acrylonitrile-butadiene-styrene (ABS). This material consists of not two but three monomers and is called a **terpolymer.** Whilst polystyrene alone is very brittle and transparent, the combination of these three monomers produces a material that is tough, opaque and glossy.

1.3.5.1 Ionomers

An interesting variation in copolymer technology is in the production of materials called ionomers. These are thermoplastic materials usually based on ethylene. However, because of their ionic chemical make-up they are able to form polar bonds (one made up of a positive force and a negative force) which lead to physical crosslinking. As you will see if you read on, this is a property usually attributed to a group of materials called the thermosets. The difference between thermoplastic and thermosetting behaviour is explained below.

Because of the polar bond, the formation of crystals is suppressed. As crystallinity causes transparency to be lost, ionomers are exceptionally transparent as well as being tough, impact resistant and scuff resistant. They can be utilised in diverse applications such as dog chews, bath and kitchen door handles, footwear and the skins of golf balls.

This material belongs to a group called the polyolefins, which are discussed in section 1.4.1.1.

1.3.6 Macromolecular behaviour

The large size of polymer molecules causes them to have some unusual properties. Consider a single monomer of ethylene and then slowly increase the number of repeat units attached. This will increase the length of the carbon backbone and therefore the polymer chain. This would result in a surprising change in physical state. For example, one monomer unit would be a gas at room temperature. By six units, the expected physical state is a liquid. By two hundred and fifty units, it would resemble a hard wax and at several thousand, it would be a hard plastic material.

These states, gas, liquid and solid, apply when considering varying lengths of polymer chain. However when considering the lengths of chain which form polymeric materials, these states do not help us explain the nature of polymeric materials. Instead, the following terms are used: thermoplastic and thermoset.

Thermoplastics solidify as they cool, and at a certain point the long molecules can no longer move freely. However, on reheating they regain the ability to flow. Hence they are frequently described as melt-processable. Thermoplastics are further subdivided into materials exhibiting amorphous or semi-crystalline behaviour (**Figure 1.22**). This distinction was touched upon already in section 1.3.2.

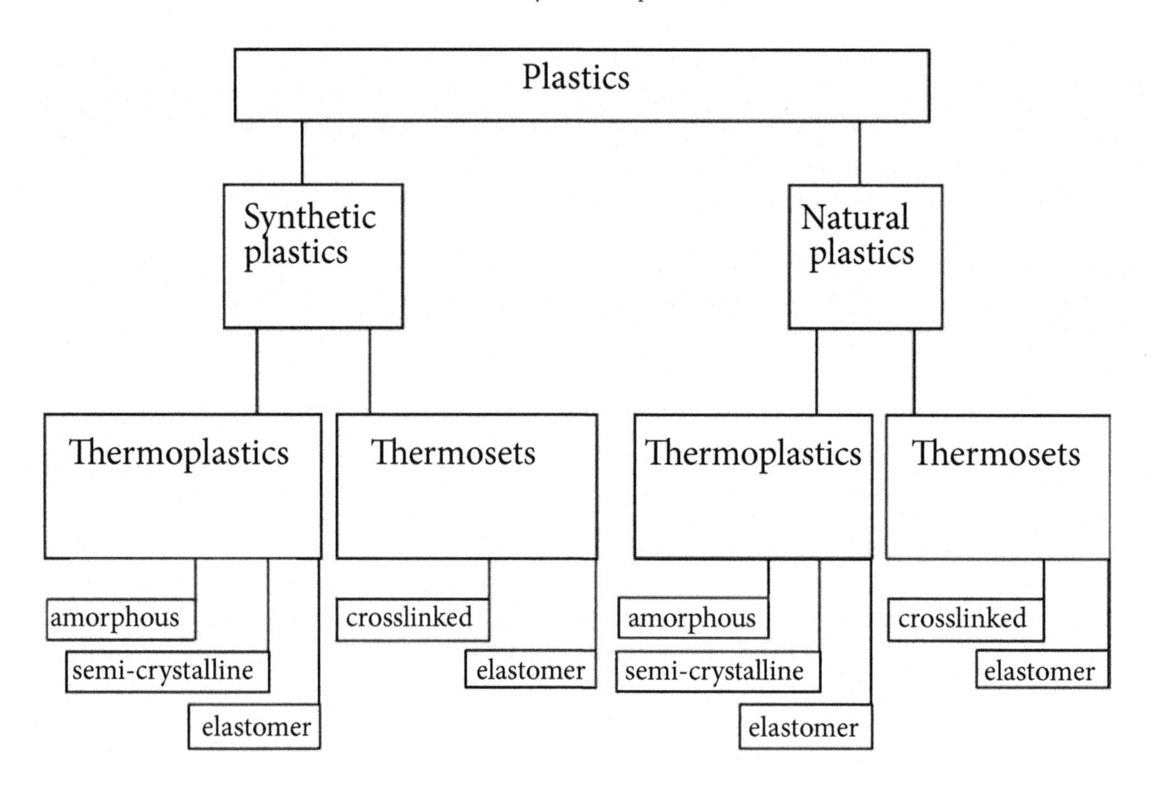

Figure 1.22

Hierarchy of plastic materials

A thermoset material solidifies by being chemically cured, creating a crosslinked structure. These molecules cannot slide past each other and the ability to flow is lost. These are high density materials which are stiff and brittle. Thermosets materials are discussed more fully in section 1.4.2.

Both thermoset and thermoplastic materials can be used to create a third set of materials called elastomers. As the name suggests this is when a material can behave very elastically like rubber, and the words elastomer and rubber are often used interchangeably. Like thermosets these materials are crosslinked but the density of the crosslinks is much lower. Elastomers will be introduced in section 1.4.3.

1.3.6.1 Amorphous materials

Amorphous materials have a totally random molecular arrangement (like a bowl of spaghetti) and soften over a broad temperature range. On cooling from a molten state the molecular chain structure remains in a random disordered state. Below a temperature known as the glass transition temperature (T_g), the strands solidify and therefore the chains do not move. The individual rotations of carbon to carbon bonds are frozen.

The random nature of the polymer chains leads to a very useful property of amorphous materials. They are generally transparent. This random arrangement means visible light is not hindered as it passes through the polymer. The largest ordered region is a carbon-carbon bond, which is very small even compared to the wavelength of visible light. This allows light waves to travel straight through the material.

These materials do not actually melt if they are heated, but they gradually become soft enough to allow the polymer to flow as the polymer chains become more mobile. (However, 'melting' points are often quoted for these materials. These figures are indicative of when the material will flow rather than a defined melting point from solid to liquid as seen in semi-crystalline materials.) The polymer is solid at certain temperatures. There is also a point at which the polymer is liquid. However, this is not a melting point of the polymer but a transition point, whereby the chains have enough energy to slide past each other. To return to the cooked spaghetti analogy, strands tend to be flexible. They twist and entangle with each other. When a polymer is molten, if you were able to grab one end of a strand, the strand would slide and eventually untangle, but if the polymer is cooled below T_g, this ability to slide is lost. (The spaghetti is dry and brittle again.) However, the glass transition temperature depends on the molecular weight. If the average length of the polymer chains is increased, these polymer chains have more difficulty sliding past each other and the temperature required to obtain flow would be higher.

In polystyrene, this glass transition temperature is around 110 °C, depending on the chain length distribution. Examples of T_g for some common amorphous materials are shown in **Table 1.6**. This table also contains examples of semi-crystalline materials such as polypropylene (PP) and polyethylene (LDPE). It can be seen that T_g is negative for both PP and LDPE but positive for PS. The implications of whether T_g is above or below room temperature can be seen in whether a material is glass-like and brittle at room temperature, like PS, or flexible at room temperature like PP and LDPE. If PP or LDPE was cooled to below its T_g it would exhibit glassy brittle behaviour similar to PS.

1.3.6.2 Semi-crystalline materials

In crystalline materials there is a clear point at which the material melts and flows, and this is the melting point (T_m).

Table 1.6
Glass transitions and melting points of common polymer materials

Polymer	Amorphous or semi-crystalline	Density (g/cm^3)	T_g (°C)	T_m (°C)	Temperature of liquid like behaviour (°C)
LDPE	Semi-crystalline	0.92	-125	135	135
PA 6	Semi-crystalline	1.13	75	233	233
PC	Amorphous	1.2	149	-	267
PET	Semi-crystalline	1.4	69	245	245
PP	Semi-crystalline	0.90	-20	170	170
PS	Amorphous	1.05	110	-	240
PVC	Amorphous	1.4	82	-	100

Crystalline behaviour depends on the ability of the polymer chains to align into ordered regions. Therefore the chain structure and the arrangement of atoms around the backbone affect the ability of materials to form crystals. Because the macromolecules are so big, if the structure is favourable for crystal formation the chains can rotate and align. However, numerous growing crystalline sites begin to form at the same time and therefore impinge on one another. It is perhaps not surprising that amorphous regions also exist where alignment is hindered. So this type of material actually contains both amorphous and crystalline regions as shown in **Figure 1.23.** 100% crystalline formation cannot be achieved and a semi-crystalline structure develops. The amorphous and crystalline region interfaces reflect light and cause these materials to appear opaque and milky.

Crystalline structure begins to develop if the material temperature is reduced below its melting point (T_m), and chains begin to move into ordered regions. Between the melting point and the glass transition temperature, the material behaves as a 'leathery' solid as the amorphous component of the polymer is still unfrozen.

The crystalline region, though comparatively small, can be 50-500 μm (μm, a micrometre, is 10^{-6} m). If the temperature is reduced further, the remaining amorphous regions will solidify at T_g. Semi-crystalline materials tend to have glass transition temperatures that are below zero. Therefore, at room temperature these materials exhibit leathery or rubbery behaviour, as opposed to amorphous materials which have brittle and glass-like properties.

The formation of these crystalline regions and their size depends on the cooling rate of the polymer. Semi-crystalline materials have very sharp melting points like the polyolefin (LDPE and PP) and nylon (PA 6) materials shown in **Table 1.6**. They also tend to shrink during cooling due to this molecular rearrangement. In some cases this can reach 20%. This shrinkage does not tend to occur evenly through a plastic component as it depends on how the chains are aligned as they cool.

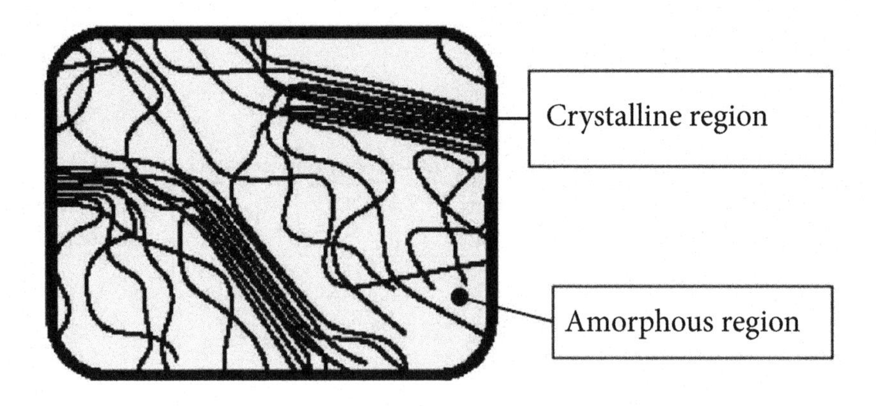

Figure 1.23

Semi-crystalline structure

Table 1.7
Semi-crystalline polymers: melting point and degree of crystallinity

Polymer	Abbreviation	T_m	Degree of crystallinity (1 is 100% crystalline)
High density polyethylene	HDPE	141	0.8
Polypropylene (isotactic)	PP	172	0.6
Poly(ethylene terephthalate)	PET	275	0.5
Nylon 66	PA 66	272	0.7
Nylon 6	PA 6	223	0.5

Not surprisingly given their differences in molecular arrangements, the different polyethylenes exhibit differing levels of crystallinity. Consider two structures of polyethylene, HDPE and LDPE, which were originally described in **Figure 1.11** and **Table 1.2**.

HDPE is a linear molecule, with a small number of short side chains. This is a favourable arrangement for chain alignment, and HDPE crystallises to about 80%.

On the other hand LDPE exhibits long chain branching. This hinders crystal formation compared with HDPE. This material crystallises to 40-50%. Examples of the degree of crystallinity achieved in some common semi-crystalline materials is shown in **Table 1.7**.

The effect of decreasing chain mobility (less crystalline formation), is a decrease in both T_m and T_g. Chain flexibility is particularly important for the level of T_g. This is because polymers capable of large scale molecular motions at very low temperatures have a low T_g. For example, the simplest structure polyethylene has a T_g of -125°C. In atactic polypropylene, where the substituent group is CH_3 rather than just H, this rises to –20°C. In PVC where the atomic constituent is chlorine, T_g is at 89°C. In polystyrene (PS), where there is an aromatic ring, this rises to 100°C. As the molecular motion is reduced, T_g rises.

The presence of crystals in the structure of semi-crystalline polymers also imparts a wide range of useful properties compared to amorphous materials with their random chain structure. These include chemical resistance, fatigue resistance and resistance to stress cracking.

1.3.6.3 Practical implications

Therefore, in conclusion to this section on amorphous and semi-crystalline materials it should be restated that the glass transition (T_g) is not the same as the melting point (T_m).

A glass transition point is a property of the amorphous region whether within an amorphous or semi-crystalline material.

The melting point is a property of a crystalline region. Below T_m, the crystals are ordered, above T_m the crystals are disordered. Below T_g, amorphous materials are disordered and immobile. Above T_g, amorphous materials are disordered but portions can move. This is illustrated in **Figure 1.24.**

An overview of some of the properties exhibited by amorphous and semi-crystalline materials is given in **Table 1.8.**

The basic behaviour of the polymers and numerous factors which affect properties have now been introduced. To see how these properties can be utilised by industry two different applications are now considered.

An example of an amorphous plastic application is in automotive lenses (**Figure 1.25**), where transparency is essential. A low coefficient of thermal expansion is also helpful in automotive applications where a number of different materials are fitted together as an assembly. Differences in

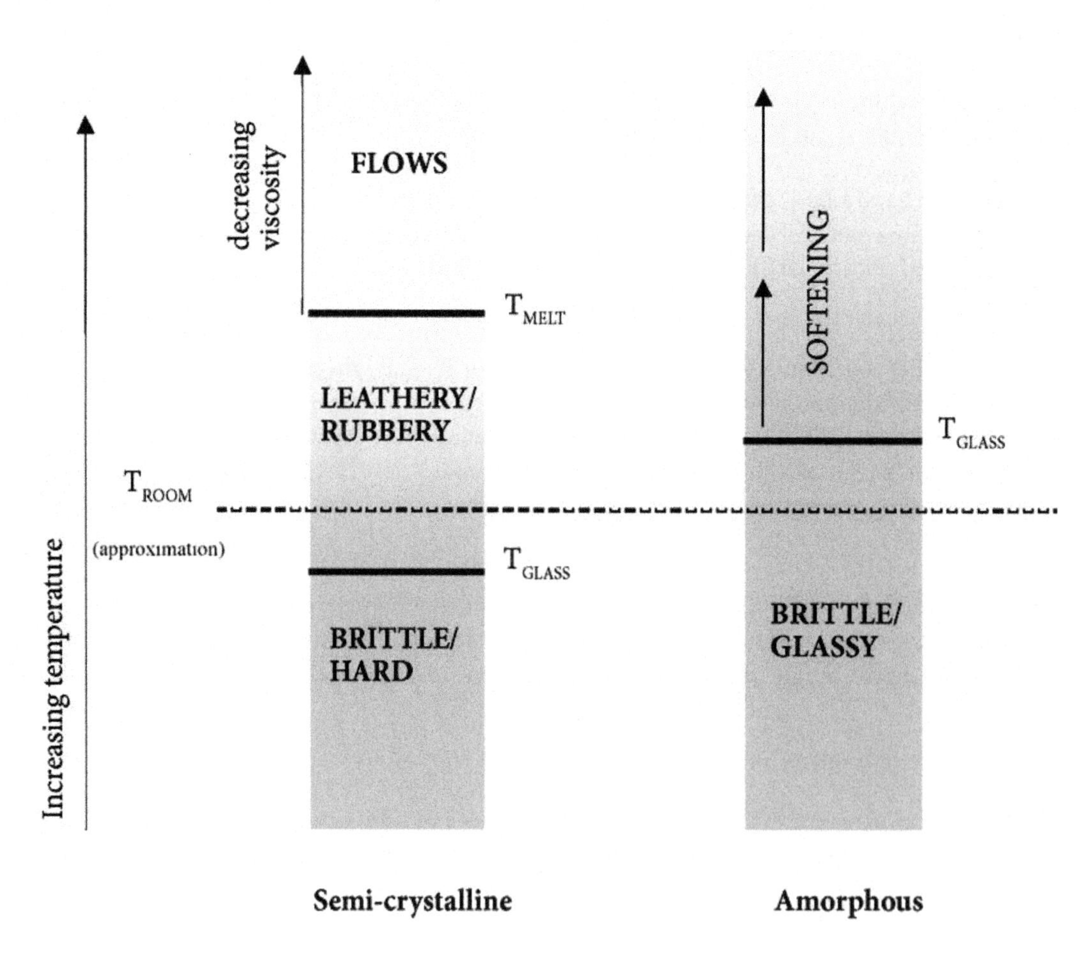

Figure 1.24

Variation in amorphous and crystalline behaviour with temperature (generalised)

coefficient of thermal expansion (CLTE) values can cause gaps or oversizing of fitted components as temperatures vary. The material properties must also be stable at operating (environmental) conditions. Polycarbonate and PMMA are both amorphous materials used in these types of automotive applications. PMMA is the more brittle, so PC is used when there is likely to be impact damage, such as in the front lenses which may be hit by a stone; however PMMA is the more naturally UV resistant of the two materials.

PC is also used in the manufacture of compact discs, eyeglass lenses, helmets, household appliances and water bottles.

A typical semi-crystalline application is in tubing. The nylons are excellent materials for this application and the most common material used is nylon 6. Nylon 66 can be used if higher performance is required.

Table 1.8
Comparison of properties of amorphous and semi-crystalline materials

Amorphous polymers	Semi-crystalline polymers
Transparent	Translucent
Low mould shrinkage (0.4-0.8%)	High mould shrinkage (1.5-3.0%)
Attacked by most solvents	Resistance to solvents
Stress cracking susceptible	Stress cracking resistant
Low coefficient of thermal expansion	Resistance to dynamic fatigue
Properties less temperature dependent, softens gradually	Temperature range increased by glass reinforcement
	Orientation gives high strength fibres
	Retention of ductility on short term heat ageing
For information on properties see Part 3	

Figure 1.25

Automotive front lens in PC (current) and glazing (future) combining the properties of both PC and PMMA

Figure 1.26

Nylon tubing

Nylon tubing (**Figure 1.26**) is lightweight, has a long fatigue life, excellent abrasion resistance, corrosion resistance and good chemical compatibility. Nylons can consequently be used in a whole range of tubing applications from pneumatics, hydraulics and vacuum applications to tubing for lubricants, fuel, petrochemicals, general chemicals, coolant gases and fresh and salt water.

Other applications for the nylon family include electrical connectors, gears, slides, cams, and bearings, cable ties, fluid reservoirs, fishing line, brush bristles, carpeting, sportswear, and sports and recreational equipment.

1.3.7 Viscosity and flow

A further property of polymers to consider is how they behave when they are molten and forced to flow. This is necessary for any consideration of the plastic forming processes used in industry. Plastics processing will be covered fully in Part 2.

The most familiar flow to consider is that of water when a tap is switched on. The pressure in the pipes forces the water to flow when the tap is opened. When the tap is closed the water does not flow.

If a molten polymer was to appear when you turned on the tap, it would not flow as easily and would appear 'thicker' (more viscous) than water, due to the entanglement of all of the polymer chains. Therefore polymers have a higher viscosity than water.

Viscosity is a term used to measure and compare the flow behaviour under an applied load (such as the pressure forcing water out of a tap). This field of study is called rheology. Therefore, the rheological properties of a polymer refer to how it behaves when it is forced to flow. To do this the polymer must be deformed; a pressure must be applied.

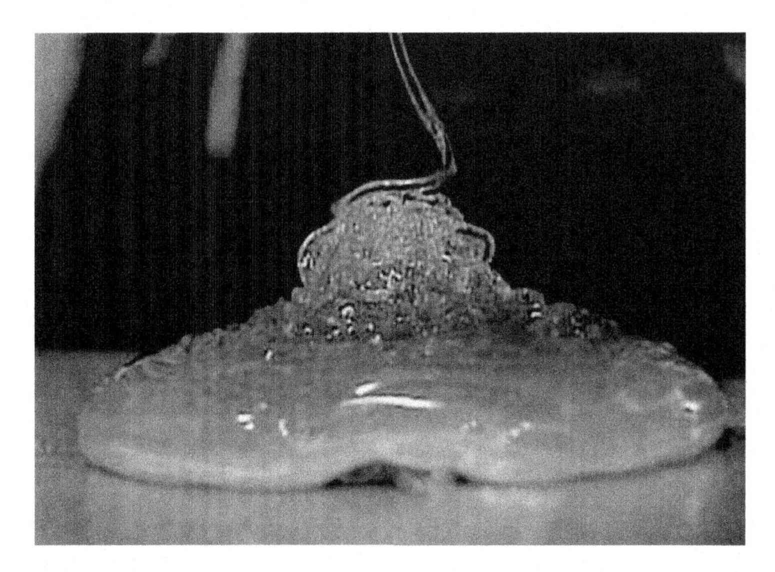

Figure 1.27

Molten nylon (type PA 6)
Source D. Bate

If you imagined stirring a bowl of water and stirring a bowl of molten polymer, generally the consistency of molten polymer is thicker than water, and this viscosity can be altered by temperature and by pressure. An example of what a molten polymer looks like is shown in **Figure 1.27**.

However with all the possible variation in polymer materials, the actual viscosity varies not just from material to material but also with factors such as molecular weight (average length of the polymer chains), chemical structure, temperature and pressure. A further factor affecting the viscosity is the speed at which the polymer is deformed.

In order for thermoplastics (synthetic or natural materials such as bioplastics) to be formed into numerous useful applications, it is necessary for them to flow. Processing plastics in its simplest form can be broken into three stages: flowing, forming and solidifying.

The ability of a plastic to flow depends on two functions of the polymer within it. These are the temperature relative to the material's T_g and the time for which the material is put under a stress to make it flow. The type of behaviour polymers exhibit is termed viscoelastic. This word suggests both types of properties the polymer can show: that of a viscous liquid or that of an elastic material.

As most plastics are not molten at room temperature, they require the input of heat in order to be processed. Once the polymer is molten, it then requires a process to deform and thereby form it. It then requires heat transfer to return the plastic to room temperature. As a consequence of these actions, there will be changes to the structure and properties of the plastic material.

To enable the polymer within the plastic to flow it must be heated above the glass transition temperature (amorphous materials) or above the melting point (semi-crystalline materials).

However, once materials are molten their flow behaviour is determined by the interaction of their viscosity and elasticity. The viscosity of a polymer is highly dependent on temperature. The elasticity of a polymer is a measure of its response to the removal of stress (for example whether it returns to its original shape – which is an elastic response). Individual polymers respond differently to both temperature and stress, and polymer melts exhibit a very wide range of viscosities from 2-300 Pa.S. (Water is 1 Pa.S.)

In industry a single point termed the melt flow index (MFI) is often used as a measure of a plastic's ability to flow at a certain temperature. This and other methods of assessing polymer flow will be introduced in Part 3 of this book.

Many useful properties of polymers can be measured: strength, stiffness, impact resistance, UV resistance, electrical conductivity and water absorption to name just a few that will be covered in Part 3. However, it is as plastic materials and not as polymers that these materials impact on our lives. Turning polymers into plastics will therefore be the focus of the next sections.

1.4 Plastics

The actual amount of polymer within a plastic material can vary widely. Some may contain virtually 100% polymer whilst in others there may be less than 10%. In industry, the very high polymer grades are often called '**prime grades**'. Making up the rest of the plastic are a variety of materials called additives. Some of these materials are functional; with some acting simply as fillers (fillers are much cheaper than polymer). There are many additives which can be added to the polymer. The final composition of the plastic, called the plastic formulation, will depend on the final intended application for the material.

Additives will be discussed more fully in section 1.5. Given the number of different polymers, copolymers, and various formulations that can be created, there are literally thousands of different grades from which to choose. This is where the inherent versatility of plastic materials can be clearly seen.

Plastics can be subdivided into three main categories, thermoplastics, thermosets and elastomers, based on the polymers they are made from. This distinction is based on both the molecular structure and the processing routes that can be applied. Examples of each group are given in **Table 1.9**.

1.4.1 Thermoplastics

About 90% of all plastics used today are thermoplastics. These materials melt and flow when they are heated, and solidify again as they cool down. This allows them to be easily formed by heat processes. They can also be re-melted and subsequently formed again, so the nature of a thermoplastic is a material that can be recycled repeatedly. However, the reality of recycling is not quite so straightforward and will be discussed in Part 4.

Thermoplastics are used to make many of our familiar consumer items such as drinks containers, carrier bags and buckets as well as a variety of other applications. The most common thermoplastic materials and their applications are shown in **Table 1.10**.

Table 1.9
Examples of thermoplastic, thermoset and elastomer materials

Thermoplastics	Thermosets	Elastomers
All plastics in **Table 1.1***	Epoxies (EP)	Diene** elastomers, e.g. polyisoprene (natural rubber), polybutadiene, EPDM****
Polycarbonate (PC)	Phenolic resins (phenol-formaldehyde, PF)	Non-diene elastomers, e.g. polyisobutylene, polysiloxanes, elastomeric polyurethane materials (commonly known as spandex or Lycra), SBS, SEBS***, EPM****
Poly(methyl methacrylate) (PMMA)	Urea resins (urea-formaldehyde, UF)	
Nylons (PA)	Polyesters (unsaturated) (UP)	
Polyacetal (POM)	Melamine resins (melamine-formaldehyde, MF)	
Poly(ether ether ketone) (PEEK)		

** Polyurethanes actually include thermoplastics, thermosets and elastomers.*
*** A diene is a polymer made from a monomer that contains two carbon-carbon double bonds.*
**** SBS refers to styrene/butylene/styrene block copolymer, (SEBS also contains ethylene).*
***** EPM is a copolymer of ethylene and propylene, (Note EPDM contains an extra diene group). The E refers to ethylene, P to propylene, D to diene and M its rubber class.*

The most widely used class of plastic is polyethylene which on its own accounts for 40% of all the thermoplastic usage (PP is the most common single material, but the PE family encompasses a number of different materials, including LLDPE, LDPE and HDPE). The thermoplastics groupings: commodity plastics, engineering plastics and specialty plastics have already been introduced in section 1.2.

As well as the individual names given to plastics, there are a number of other terms which are commonly used to describe a family of similar plastic materials. The words polyolefins and nylons both cover a number of different but related materials with different properties.

1.4.1.1 Polyolefins

In section 1.3.2 it was shown that polyethylene was formed during breakage of linear C=C double bonds. All polymers made with similar monomers are termed polyolefins. A polyolefin can be defined as produced from a simple olefin as a monomer. Outside of the petrochemical industry, these olefins are more likely to be called alkenes.

This family includes all variations of materials with different levels of branching which in turn produce different polymer characteristics. In the case of polyethylene this gives rise to the three most commonly used types of polyethylene; high density polyethylene (HDPE), low density polyethylene (LDPE) and linear low density polyethylene (LLDPE).

Table 1.10
Common thermoplastics and their applications

Thermoplastic polymer	Applications	Interesting fact
High density polyethylene (HDPE)	Packaging, pipes, toys, detergent bottles, crates, bags	The first major use of this material was in the 1950s for making hula hoop toys
Low density polyethylene (LDPE)	Packaging, grocery bags, toys, lids, squeezy bottles, pallet film	In 1935, ICI Alkali Division laboratories produced a material known as polyethylene or polythene. This material had dielectric properties which were to be vital to wartime development of radar. This material subsequently became known as LDPE as the other types of PE were developed
Linear low density polyethylene (LLDPE)	Packaging, similar markets to LDPE	In the 1970s the so-called metallocene catalysts produced controlled amounts of short chain branching to give products such as LLDPE. These can be can now be produced by other catalyst routes
Polypropylene (PP)	Caps, yoghurt pots, suitcases, tubes, buckets, rugs, battery casings, ropes, car bumpers	The most popular plastic on the planet
Polystyrene (PS)	Mass produced transparent articles, yoghurt pots, fast food foamed packaging (styrofoam)	This crystal clear and brittle material can be transformed to a high impact material by the inclusion of rubber in its structure to form HIPS (high impact polystyrene) and ABS
Nylon (PA)	Bearings, bolts, skate wheels, fishing lines, carpets, clothing	DuPont developed this material in the 1930s. By 1940, the whole of the USA was supplied with nylon stockings, such was the impact of this material – see **Figure 1.28**. Also known as nylon
Poly(ethylene terephthalate) (PET)	Transparent carbonated drink bottles, film, audio and video tapes	The first PET bottle was introduced in 1979. Fleece clothing can be made from recycled PET bottles
Poly(vinyl chloride) (PVC) PVC-U (unplasticised) PVC-P (plasticised)	Comes in both flexible and rigid forms PVC-U :Food packaging, Window frames, drain pipes, foams PVC-P: flooring, rainwear, foams, tablecloths	PVC-U: Second largest volume plastic on the planet after PP PVP-P: Flame and water resistance made it critical for wire insulation during World War 2, also known as vinyl

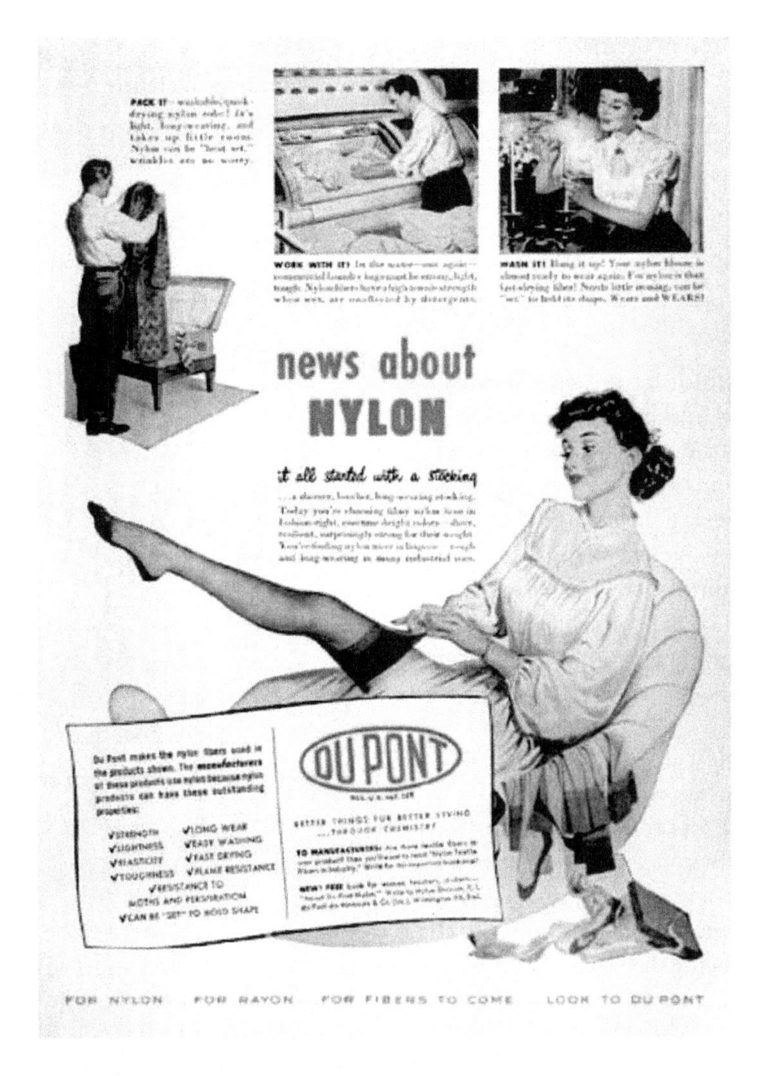

Figure 1.28

1948 advert for DuPont nylon

Picture courtesy of DuPont

Differences in the polymer chain result in a number of material variations such as in melting points, stiffness and density. This makes them all suitable for slightly different applications. It is therefore also necessary to identify the type of polyolefin being referred to. The polyolefin family of materials also includes polypropylene (PP), another commonly used material. Note, all the densities of these polyolefins shown in **Table 1.11** are below the density of water. These materials will float.

1.4.1.2 Polyamides or nylons

This group of materials are more commonly known as nylons. DuPont who commercially developed the material in the 1930s gave it this generic name which is still widely in use today. These polymers

Table 1.11
Melting points and densities of common polyolefins

Polyolefin	Melting point (°C)	Density (g/cm^3)
LDPE	115	0.92
LLDPE	123	0.92
HDPE	130	0.95
PP	170	0.90

all contain the amide group (a chemical group of molecules as shown in **Figure 1.29**) so are also called the polyamides. Like the polyolefins, there are a number of ways the multiple monomers can be arranged but also different nylons are polymerised from different monomers. This difference gives rise to the numbering system of the nylons. Commonly seen examples are nylon 6 and nylon 66, these materials may also be written as PA 6 or PA 66.

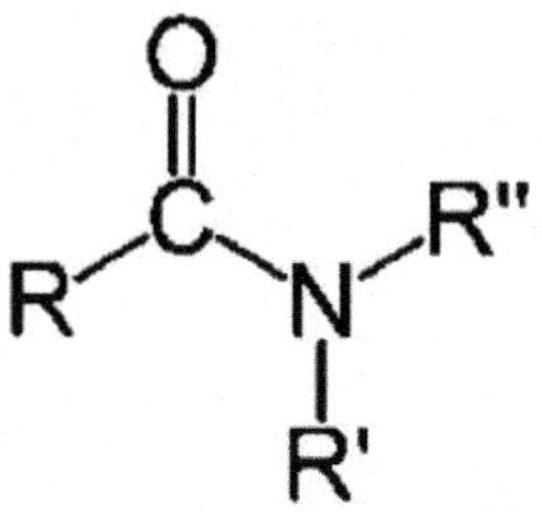

Figure 1.29
Chemical representation of an amide group

There are a whole range of other nylons available which are less widely used than PA 6 and PA 66, such as PA 11 and PA 12. Again, like the polyolefins, the properties of the various types of polymer can vary greatly due to these differences in the make-up of the polymer chain. Therefore, it is important to be clear on the type of nylon material being discussed. Examples of the properties of the four most common nylons are given in **Table 1.12**.

The numbering system of the nylons relates to their molecular structure and the number of carbon atoms linked together in the various repeating units. However, a more detailed explanation is beyond the scope of this book.

A further point to note is the differences in properties between the polyolefins and the nylons. Quite a range of densities and melting points are exhibited in these two groups of materials.

Table 1.12
Melting points and densities of common nylons

Nylon	Melting point (°C)	Density (g/cm^3)
Nylon 6	233	1.13
Nylon 66	265	1.14
Nylon 11	180	1.04
Nylon 12	180	1.02

1.4.1.3 Other thermoplastics

We have already seen that thermoplastic materials can be split into two types amorphous and semi-crystalline. The lists provided in **Tables 1.13** and **1.14** are not exhaustive and there are many other materials available.

Table 1.13
Examples of amorphous materials

Material		Typical properties	Examples of use
Commodity materials			
PS	Polystyrene	Transparent and brittle	Disposable packaging, insulating film, toys, drawing instruments, disposable cutlery, clock cases
PVC Available in two types: unplasticised and plasticised, giving a rigid or flexible material choice	Poly(vinyl chloride)	PVC-U: strong, rigid, tough. Good chemical resistance PVC-P flexible, toughness is temperature dependent	PVC-U: window frames, pipes, disposable packaging and foams PVC-P: handles, plugs, protection caps, shock absorbers, foams
Engineering materials			
ABS	Acrylonitrile-butadiene-styrene	Tough, stiff, abrasion resistant	Vacuum cleaner housings, telephone handsets, mobile phone cases
PSU	Polysulfone	Stiff, strong, good dimensional stability, transparent	Passenger overhead service units in aircraft, high temperature electrical components
PC	Polycarbonate	Tough, stiff, strong, transparent, good electrical properties	Helmets (safety, motorcycle), street lamp covers, feeding bottles (babies)
PPE (modified)	Poly(phenylene ether) – modified (This is a blend of PPE and PS in the ratio 1:1)	Low density, high strength and stiffness, low water absorption, good electrical properties	Parts for cameras, televisions, radios, electric kettles, hair dryers, office equipment
PMMA	Poly(methyl methacrylate) Also known as acrylic	Hard, rigid, weather resistant, transparent	Outdoor signs, glazing, aircraft canopies, skylights, auto tail lights, machine covers
SAN	Styrene-acrylonitrile copolymer	Rigid, hard and scratch resistant	Packaging for food, pharmaceuticals and cosmetics, covers, inspection ports, coffee filters, triangular warning signs
High performance materials			
PES	Poly(ethersulfone)	High strength, high thermal stability	Coil cores, carburettor components, lamp holders, spotlights, lenses, aircraft nose cones
PEI	Poly(etherimide)	Very high strength, stiffness, hardness, heat resistance and weathering resistance. High dielectric strength	High voltage circuit breaker housing, transmission components, safety belt clips, brake cylinder parts

Table 1.14
Examples of semi-crystalline materials

Material		Typical properties	Examples of use
Commodity materials			
PP	Polypropylene	Lightweight, low cost	Disposable packaging
LDPE	Low density polyethylene	Lightweight, low cost	Disposable packaging
HDPE	High density polyethylene	Lightweight, low cost	Disposable packaging
Engineering materials			
PA 6 and 66	Nylon 6 and 66	High strength and stiffness. Tough, resistant to heat, abrasion and most solvents. These materials absorb water, which must be taken into account in designing. PA 6 is cheaper but PA 66 has a higher melting point, lower creep and greater stiffness	Gears, bearings
PBT	Poly(butylene terephthalate)	Excellent dimensional stability, high strength and stiffness, low creep, excellent electrical properties, good chemical resistance, low mould shrinkage	Appliances, automotive, electrical and electronic consumer products
PET	Poly(ethylene terephthalate)	High strength, stiffness and dimensional stability, good slip, wear and electrical properties	Wear resistant high precision parts: gears, bearings, cams, guides. Clear bottles, insulating film, release films, magnetic tape
TPE	Thermoplastic elastomers	Rubber-like properties	Tubes and hoses, shoes
POM	Polyacetal	High toughness, stiffness, hardness, dimensional stability, resistant to solvents, stress cracking	For precision engineering applications, plumbing, car parts, domestic appliance housings
High performance materials			
LCP	Liquid crystal polymers	Flame retardant, resistant to organic solvents, very high mechanical properties (but generally anisotropic)	Aviation, components for electronics such as connectors, sockets, chip carriers
PTFE	Poly(tetra-fluoroethylene)	Not used for mechanical properties, however excellent chemical, electrical and thermal properties. Insoluble, high thermal stability, low coefficient of friction	The non stick surface on frying pans is PTFE. (Teflon is one well known trade name of this material). This kind of repellent coating is a typical application, PTFE is also used for packaging, seals, piston rings, tubes, hoses and fittings
PPS	Poly(phenylene-sulfide)	High strength, stiffness, hardness, low moisture absorption, high dimensional stability, chemical resistance and weathering	Ball valves, pump housings, impellers, electrical components
PEEK	Poly(ether-etherketone)	Strong and tough. Good fatigue and heat resistance, chemical resistant	Cable sheathing, high performance components for automotive, aircraft and electrical industry
PI	Polyimide	High heat resistance	Piston rings, valve seats, coil cores, compressor seals in jet engines

1.4.2 Thermosets

Whilst the majority of plastic in use is thermoplastic, the remaining 10% is made up of materials called thermosets. Unlike the thermoplastics they cannot be melted, solidified and then melted again. The thermosets undergo a process called crosslinking during polymerisation. This causes the material to become resistant to heat. At very high temperatures these materials will decompose rather than melt. Thermosets are initially melt processed in a similar manner to the thermoplastics, once formed they cannot be reprocessed.

This chemical crosslinking process is termed 'curing' and results in a highly dense molecular network that makes the material stiff and brittle. The differences in the arrangement of molecules between thermoplastics and thermosets can be seen in **Figure 1.30**.

This crosslinking, like the polymerisation reaction used to produce polyethylene chains, is usually the result of the breaking of double bonds between carbons, allowing linkages to form. However in the case of thermosets, there is usually a by-product. In the case of a phenol-formaldehyde polymer (often just referred to as phenolic) this by-product is water. This material is the most common thermoset, around 42% of usage is phenolics.

Common thermosets and their applications and market share are shown in **Table 1.15**. Thermosets are often used where their strength and durability can be utilised. These materials may be used in very demanding applications for many years. This strength and durability depends on the arrangement and frequency of the chemical crosslinks; in the thermosets described above they are close together. In materials where the crosslinks are widely spaced, this produces elastic rubber-like behaviour. Such thermosets are called elastomers and natural rubber is an example of this. There are also thermoplastic elastomeric materials available.

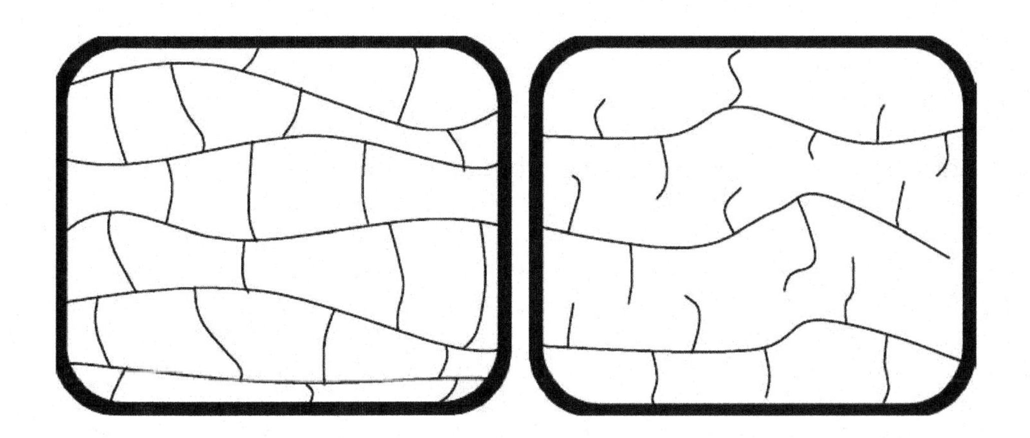

Figure 1.30

Highly crosslinked structure and uncrosslinked structure. Thermoset crosslinks (left), thermoplastic chains (right)

Table 1.15
Common thermoset materials

Thermoset polymer	Application	Thermoset market share
Allyls (DAP and DAIP)	Commercial and military electrical connectors, circuit breaker housings, X-ray tube housings, dental equipment	Small (not included in total % share)
Aminos (melamine-formaldehyde and urea-formaldehyde)	Melamines: scratch resistant laminate surfaces, e.g. kitchen worktops, tableware and picnic ware Ureas: closures, control housings, wiring devices, control buttons, knobs	4% melamine and 27% Urea
Epoxies	Adhesives, electrical insulation, composites	8%
Phenolics (phenol-formaldehyde)	Items requiring resistance to heat and chemicals. Heat resistant handles for pans, irons, toasters, circuit boards, fibreglass binder, plywood, electrical meter casings	42%
Thermoset polyesters	Automotive components, brush holders, battery racks, toaster sides, partitions	19%
Silicones are also thermoset materials but they are beyond the scope of this book. They are inorganic materials i.e. they do not contain a carbon backbone but an alternating silicone and oxygen one.		

Because of their high performance, thermoset materials are often hidden from view and therefore less familiar to the layman than more common thermoplastic materials. However where there are high performance operating environments, there are most likely thermoset materials. As one example such a component may be hidden in your computer withstanding high temperatures. A further component may be used in aerospace applications withstanding centrifugal forces of thousands of revolutions per minute (RPM) or within a chemical engineering environment which requires considerable corrosion resistance at elevated temperatures and pressures. Thermoset materials are also often used in paints and coatings where their chemical resistance and durability can be used to best advantage.

The chemistry of thermoset formation is far more complex than that of thermoplastics and beyond the scope of this book. However as a thermoset reacts and crosslinks, there is a change in viscosity and this effect is shown in **Figure 1.31**. As thermoset polymers are very hard and brittle, additives are used to alter the material properties. Thermoset plastics (as opposed to thermoset polymers) are composed of a number of ingredients such as the polymer (frequently called a resin), reinforcements and/or fillers, pigments and/or dyes, catalysts, lubricants and solvents.

The thermoset material may be formulated to optimise flow, electrical insulation, mechanical strength, chemical resistance, flame resistance, surface finish, colour or weather resistance. Typical applications are shown in **Figure 1.32**.

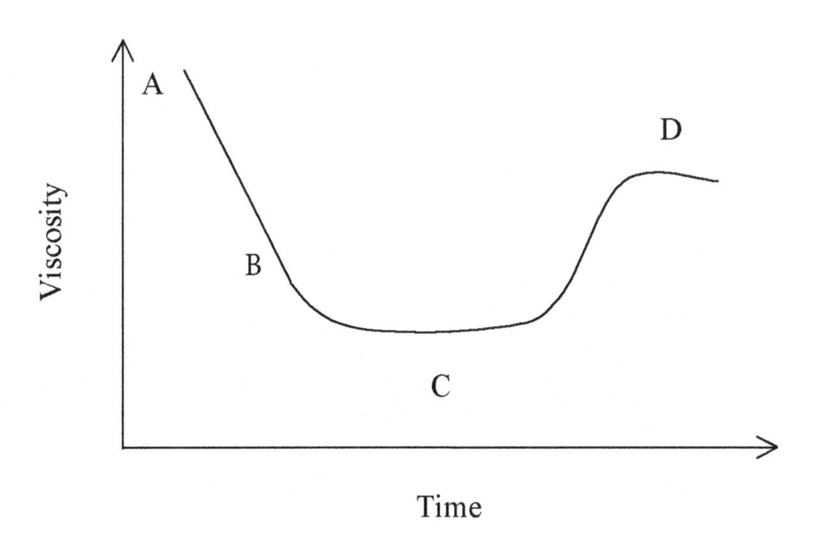

Figure 1.31

Progress of viscosity in a typical thermosetting reaction: A - Reaction begins (highest viscosity point); B - Mix increases in temperature leading to rapid drop in viscosity; C - Viscosity decreases to minimum before rapid acceleration of reaction from C to D; D - Reaction ceases or cure completes (whichever occurs first)

Figure 1.32

Typical applications of thermosets: phenolic electrical meter casing knob assembly (left), melamine plates (right)

The thermoset resin needs to be tested for viscosity, gel time (the time in which the material melts and flows but is not yet crosslinked), cure rate and solubility. This is to ensure that the optimum cure properties and performance are achieved. Until it is crosslinked it can be melted and remelted just like a thermoplastic. However once cured and crosslinked it cannot be melted again.

1.4.3 Elastomers or rubbers

Like thermosets, elastomers are crosslinked but they form coarse meshed macromolecules rather than a dense network of crosslinks.

A polymeric elastomer can be defined by its ability to stretch repeatedly to twice its original length at room temperature. Elastomers are very useful materials where their flexibility and ability to regain their shape makes them ideal for automotive, medical and sports applications (**Figure 1.33**). In cars they are used in applications such as sealing rings, door and window seals, car wipers, and soft-grip handles.

These materials can be thermoset materials such as polyisoprene (natural rubber) or thermoplastic materials called thermoplastic elastomers (TPEs). TPEs are gaining in popularity as they not only have elastic properties like thermosetting rubbers but they can also be reheated and reprocessed like standard thermoplastics on conventional processing machinery. One example, the copolymer SEBS was shown in **Table 1.5**. This is a thermoplastic elastomer. A further advantage of these materials over thermosetting rubber is that they can be recycled more easily. They are also widely used as additives to modify the properties of more rigid thermoplastics. This increases the impact strength because of their inherent molecular structure; they form only weak bonds when they cool and solidify.

There are six main generic classes of TPEs commercially available:

- styrenic block copolymers, (TPE)
- polyolefin (olefinic) blends (TPO)
- elastomeric alloys (TPV)
- thermoplastic polyurethane elastomers (TPU)
- thermoplastic copolyesters (COPE)
- thermoplastic polyamides (PEBA)

Newer materials introduced in this sector as low-cost rubber replacements for non-demanding applications and property modifiers include:

- reactor TPOs (R-TPOs)
- polyolefin plastomers (POPs) e.g. ethylene-octene (<20% wt octane comonomer). POPs are copolymers of ethylene and an olefin such as butene or octene. They are called plastomers because of their ability to modify properties when used with other plastics, like the use of an additive. (These materials were invented by The Dow Chemical Company and sold under the trade name Affinity™.)
- polyolefin elastomers (POEs) – as POP but with >20% wt octene comonomer

Both POP and POE materials are used as modifiers in other products including materials such as TPE.

Further applications of TPEs include swimming goggle seals and straps, fins, snorkels, handheld computer housings, computer and audio connectors, toys (e.g. flying rings and discs, 'rubber' balls,

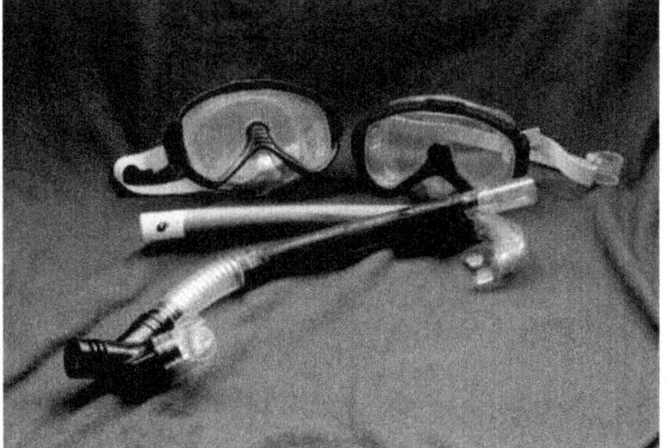

Figure 1.33

Uses of thermoplastic elastomers: soft feel eye pieces (left), goggle visor surround and straps (right)

doll body parts), automotive cup holders and gear shifter boots, catheters (nylon block copolymers), shoe soles and adhesives. COPE materials are used for snowmobile tracks.

1.5 Additives – from polymers to plastics

How do you turn a polymer into a plastic?

The answer is you modify it with one or more additives.

As well as there being numerous polymer materials, there are also many additives available to modify their properties. There are therefore literally thousands of variations of materials that can be produced. An estimate some years ago put commercial material variations on the market at 18,000 in the USA, 10,000 in Japan and 6,000 in Europe. This certainly gives a designer plenty of opportunity to find a material to suit their application. With this kind of choice it is easy to see why plastics as materials are so successful in virtually every market worldwide.

These additives take many forms and they can be added into the polymer at a range of quantities. For example, one additive may be added at just 0.005%, a different additive may be added at 85% (leaving a plastic consisting of just 15% polymer). The addition rate depends on the type of additive used and the performance level of the plastic that is required.

Additives can be simply divided into three main groups:

- Functional additives
- Fillers
- Reinforcements

1.5.1 Functional additives

As the name implies functional additives are used to impart particular properties to a plastic. There are numerous additives available in this group, some typical functions include:

- stabilisers (for adding resistance to heat and light)
- anti-static properties (e.g. to reduce dust build up on a product on a supermarket shelf)
- improving flame retardancy (e.g. increase fire resistance, reduce hot drips under fire)
- foaming agents (e.g. to foam polystyrene for packaging use)
- lubricants (e.g. to reduce friction during moulding operations)
- plasticisers (e.g. to reduce the brittleness of polymers by lowering the glass transition temperature)
- slip agents, (e.g. help with the opening of film bags)
- aesthetic purposes such as a colourant (e.g. black, white, red, metallic, special effect)
- antimicrobial (e.g. to reduce bacterial growth on a food contact surface)

1.5.2 Fillers

Fillers tend to be added to reduce cost, add volume, increase weight or improve technical performance. Depending on this purpose they are either inactive (used to reduce cost, add volume) or active fillers (serving a function). However it should be noted that the process of modifying the polymer with an active filler often presents a cost in itself, therefore an enhanced material is often as expensive as or more expensive than the polymer on its own.

There are a number of materials that are classified as fillers. There are mineral fillers such as the carbonates (calcium carbonate accounts for 70% of all carbonates), sulfates (calcium sulfate – also called gypsum, barium sulfate-blanc fixe) and silicates (mica, talc, kaolin and clay). Other filler materials include carbon black, carbon fibres, glass fibres and beads, ceramic fibres and beads, powders of metal, metal oxides and hydroxides, flours and natural fibres. All these materials can be added to polymers in various quantities and in conjunction with other additives.

Examples of common inactive filler materials used are materials such as calcium carbonate (chalk), mica, clay and talc. Whilst these materials also change the properties of the plastics, in this case they are added for the cost saving in employing them, rather than to impart any particular set of properties. Calcium carbonate is most commonly used as low cost filler in PVC, but it is also used in other plastics such as PE, PP, PU, and UP.

Natural fibres can also be used as fillers. These are plant based fibres such as hemp (**Figure 1.34**), jute, flax and kenaf. Biopolymers with biofillers make low environmental impact biocomposites. These can be composted under suitable conditions at the end of their useful life. Biopolymers will be discussed further in section 1.9.

Figure 1.34

Hemp mat, a natural fibrous filler

1.5.3 Active fillers and reinforcements

These materials are added to increase properties such as strength and hardness, e.g. short and long glass fibre, glass beads, carbon fibre and graphite.

Graphite is added as an active filler to increase electrical and thermal conductivity. It also has lubricating properties. An example of this use is with Nylon 6 in bearing bushes and similar load-bearing applications.

1.5.4 Glass fibre

In terms of volume, chopped strand glass fibre is the most important of all reinforcing (therefore active) fillers. Glass fibres find use in all kinds of engineering applications, ranging from automotive, marine and aviation to household appliances.

Glass fibres incorporated into a plastic formulation can increase rigidity, strength, impact resistance, and dimensional stability to name only four benefits. However there are disadvantages, such as low surface quality, warpage, increased abrasiveness and a higher price.

Glass fibres are produced by melting sand with various fluxes and stabilisers. The molten glass is drawn into thin fibres. They are coated in a thin layer (mostly made up of low molecular weight polymer with other additives), to protect them from abrasion and sticking together, as they are brittle and susceptible to damage. This surface treatment of glass fibres is called sizing and it produces a thin layer on the surface of the glass (20 nm to 100 nm); it also contains a coupling agent, usually a silane.

Coupling agents enable chemically dissimilar materials (glass and polymer) to stick together. This is important to ensure that maximum strength and stiffness are achieved. There are therefore specific coupling agents for specific polymer types, the commonest being silane. The size is also designed to be material specific and generally dissolves into the plastic material once the polymer-glass bonding begins.

Bundles of strands can then be prepared in a variety of ways. They can be chopped into various lengths, typically 3, 6, 12, 25 or 50 mm, or used to produce continuous filament mats and woven and non-crimp fabrics.

The use of glass mats with thermoset resins can produce very high strength and durable materials. Fibreglass is a well-known example, and materials of this type, called composites, are discussed in section 1.7.

There are three basic glass mat types:

- Unidirectional long glass fibre mats which add directional stiffness and strength along a single axis. This results in materials with high modulus, fatigue and creep resistance parallel to the unidirectional fibres (see Part 3).
- Continuous strand. These are randomly orientated glass-mat products and produce a balance of stiffness and strength in all three directional axes.
- Long, chopped fibre glass mats. These materials provide improved flow, allowing for improved energy management. There is a small decrease in stiffness, but they can be processed at low pressure.

Different grades of glass are used within the plastics industry. E Glass is the most common. Higher cost materials are available such as S Glass (higher strength, stiffness and cost). An overview of the forms of various glass fibre products can be seen in **Table 1.16**, with some examples in **Figure 1.35**.

Table 1.16
Different glass fibre reinforcement materials, showing direction(s) of fibre alignment

Reinforcement	Description	Orientation
Glass strand	Long strands	⟷
Chopped glass strand	Shorter strands	✳
Glass mat	Random directional	✳
Glass roving cloth	Interlocked two directional	✛
Glass filament cloth	Interlocked two directional	✛
Unidirectional (UD) weave	One direction of alignment	⟷

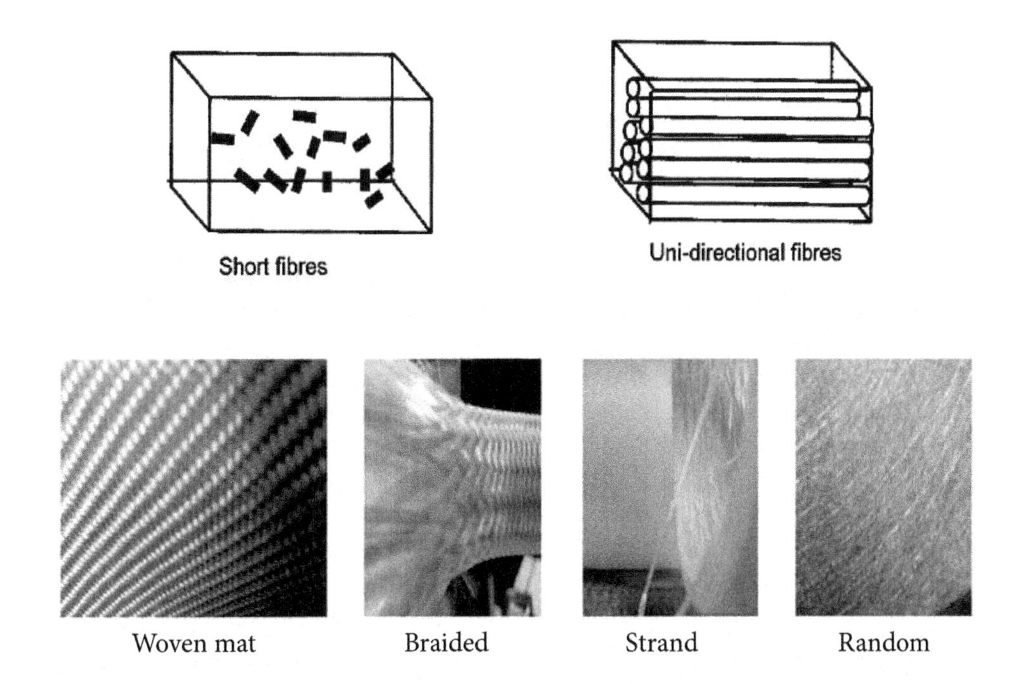

Figure 1.35

Examples of glass fibre reinforcements

1.5.5 Carbon fibre

This is an impressive material with the highest stiffness to weight ratio of any structural material. It has excellent strength, chemical resistance, and electrical and thermal conductivity. Carbon fibres also have an unusual and useful coefficient of thermal expansion in the fibre direction. This is negative and can therefore be used to compensate for shrinkage.

Carbon fibres are used in aerospace and marine applications, sporting goods, bearings and brakes. Chopped fibres between 0.5 and 0.6 mm are sized in a manner similar to glass to enhance dispersion and bonding to the polymer.

1.5.6 Others

Other less used reinforcements include aramids (one such material is more commonly known as Kevlar) and boron.

Boron fibre has similar properties to carbon fibre but is more expensive and therefore has less use. It was one of the first advanced metal fibres developed in the 1960s but today is of more historical than commercial importance.

1.5.6.1 Kevlar (Aramid)

Kevlar on the other hand is an important (and interesting) material. It is an aromatic polyamide poly(m-phenylene isophthalamide) – this group are generically called the aramids. It is known for its usage in bullet-proof vests and flak jackets woven from cloth. It is also perhaps the best known of a group of materials with special properties called the liquid crystal polymers (LCP).

Kevlar is enormously strong, for example the tensile strength of a Kevlar cable is greater than that of a steel cable of the same diameter. This amazing strength comes from the ability of the molecules within its structure to align and attain near perfect crystallinity.

Aramid fibres have ductile behaviour; they are very resistant to damage, fatigue, stress rupture and shattering on impact. However aramid fibres are expensive and are therefore generally only used in high performance applications; asbestos replacement materials (brake linings, gaskets), automotive, aircraft missile covers and armour. High quality motorcycle safety equipment such as jackets, trousers and gloves also incorporate Kevlar. Kevlar is also used in canoes, kayaks and sailboats. Lifeboats need to withstand extreme weather conditions and go out as other seacraft come in for safety. These boats need to extremely safe and therefore make use of the properties of Kevlar in hull design. The lifeboat shown in **Figure 1.36** is such a craft and is made of Kevlar and other materials called composites which are discussed in section 1.7.

These boat hulls are constructed of glass-reinforced thermoset plastic which has a foam core to give a structure over 100 mm thick. Prepreg epoxy (see section 1.7.1.4) is used with glass and Kevlar reinforcement.

Figure 1.36

Lifeboat

Source S. Hall

1.5.6.2 Nanofillers

Further examples of active fillers are nanofillers. These are materials with an upper particle size less than 100 nm. Examples include silicate nanoparticles and needle shaped nanowhiskers. Nanomaterials provide an alternative set of properties beyond the scope of this book; however nanocomposites will be briefly discussed in the composites section 1.7.2.

Examples of naturally occurring nanostructures include bone and shell. The advantages of engineering at such small scales are to impart added stiffness, toughness and dimensional stability. This is therefore an area of both scientific and commercial interest at present.

It should be noted that the use of one type of additives does not prevent the use of another, and generally a plastic may be made of a combination of materials.

For example:

- Polymer, heat and light stabiliser, a colourant, dispersion aid, chalk filler.
- Polymer, glass fibre, lubricant, colourant, heat stabiliser.

In this way the functionality and appearance of the plastic can be precisely controlled.

1.6 Blends

In some cases, the properties of a certain polymer can be enhanced by blending it with another polymer. This is also a good way of improving the cost-performance ratio of a commercial plastic. (This is very different to the production of copolymers which uses different monomers. Blending is a physical mixing of plastics already produced.) A common system for describing blends is made up of the separate polymers separated by a plus sign as seen in **Table 1.17**.

Table 1.17
Examples of commercial polymer blends

Composition	Why blend? (advantages over single materials)
(PBT + PC)	High heat distortion temperature and low temperature impact resistance
(PC + ABS)	Properties lie between components depending on blend, high impact strength over wide temperature range
(PBT + PET)	Good dimensional stability at high temperature, improved impact strength
(PP + EPDM)	High stiffness, high softening temperature, ability to modify crystallinity, low raw material cost compared with EDPM
(PPO + HIPS)	Decreased cost, increased impact strength, reduced processing temperature
(POM + PU)	High energy absorption, good elastic recovery

1.7 Composites and laminates

Plastic composite materials are increasingly being used as replacements for metallic components. They can attain similar properties whilst generally being lighter in weight than their metallic counterparts. The shift to a composite part may also enable manufacturers to integrate several previously metallic parts into one component if the design allows, often with the option of incorporating added design features. Lower processing temperatures and shorter production timescales also add to the attractiveness of a plastic composite over metals.

A composite material is defined as a solid product consisting of two or more distinct phases, including a binding material (matrix) and a fibrous or particulate material. One phase is distributed within the other. Generally this term is used to imply the use of fibres within a plastic; however the term 'composite' also covers other reinforcement and filler types. The type of reinforcement and polymer, the amount of the components, and bond between glass and polymer, all affect the mechanical properties.

Paper and concrete are also examples of composite materials. The combined properties of a composite are generally quite different from those of the constituent materials on their own. Like polymer materials, composites occur in nature (wood, bone, teeth, and insect shells) and can also be made from natural materials.

Fibreglass is probably the most familiar form of a plastic composite and has found use in a variety of applications in the automotive, construction and marine industries. This material is a thermoset plastic consisting of an unsaturated polyester matrix with glass fibre reinforcement. However, thermoplastic composites are also commonly produced. For example glass filled (15-30%) nylons (PA 6 and 66) are used for the body components of power tools, and lifting the bonnet of a car will often reveal a number of other such composite materials.

Therefore, a plastic composite can be a thermoset or a thermoplastic material. The most common composite materials are usually reinforced with glass. This is primarily due to price, as stronger materials such as carbon fibres and aramid fibres (see additives) are more expensive.

1.7.1 Examples of composite materials

1.7.1.1 GMT (glass mat reinforced thermoplastic)

GMT is a thermoplastic composite material. The material consists of a series of glass mats impregnated with the thermoplastic matrix. The glass content varies usually in the region of 30-40%. The fibres are not orientated in the mat, resulting in isotropic mechanical properties. (Isotropic properties are unrelated to orientation, whereas anisotropic properties are directional.) The GMT material is supplied as a preformed sheet or roll. This is usually polypropylene but can also be made of other polymers such as poly(butylene terephthalate) (PBT) for instance, however other materials are not widely utilised. GMT materials are processed by a technique called compression moulding (see Part 3).

The bulk of GMT is used by the automotive industry although this material has also been successfully used to produce components such as guitar cases, chairs, snow boards and boxes. However, the main disadvantage of thermoplastic composites compared to thermoset composites is generally the break-even point, i.e. the volume that must be manufactured in order to be profitable. Therefore, materials such as this are becoming less popular in the marketplace.

1.7.1.2 SMC (sheet moulding compound)

This is generally a sheet of compounded polyester-based thermoset and filler containing anywhere from 20-30% short glass fibres 20-50 mm in length, randomly orientated in the plane of the sheet. Its uncured structure means it can be handled and cut prior to heat cure. SMC is also less commonly available in higher performance grades of plastic. Grades with aligned fibres are also available. Traditionally this material is shaped by compression moulding. (See Part 3.)

1.7.1.3 BMC (bulk moulding compound): also known as DMC (dough moulding compound)

This material has the form of a dough, and it has better flow properties than SMC because it has a lower glass content of shorter length. It consists of catalysed resin, filler and 15-25% glass fibre 3-12 mm in length. Again like SMC, it is typically utilised by compression moulding but can also be processed by injection moulding for more intricate designs.

1.7.1.4 Preimpregnated glass fibre sheet (prepreg)

These materials are glass sheets already prepared with catalysed resin and therefore ready to mould under heat. The usual resin is an epoxy, with sheets about 1 mm thick of unidirectional fibres or fabric. The desired thickness is achieved by laminating a number of sheets together. This allows specific alignments of glass fibre to be achieved with a high glass content of 50-80%. These are utilised where high engineering performance is required.

1.7.1.5 Laminates

A composite can have either isotropic or anisotropic behaviour, and this depends on how it was produced. However, a composite can be considered to act as a single material with the same properties throughout its structure. A laminate by contrast, consists of distinct material layers which are bonded together, see **Figure 1.37** for a typical laminate structure.

Laminates are often produced as plastic sheeting where each layer serves a purpose. For example the outer layer may give weather resistance or scratch resistance to the layer beneath. The middle layer may provide the colour and the bottom layer may have adhesive properties to allow it to stick when in use. A commonly used laminate material in the packaging industry combines plastic and metal foil. Multi-layer packaging bottles or film can also contain a number of thin plastic layers to combine various physical properties within one material. These layers may be adhesives, barrier layers, heat resistant layers, colour layers or layers containing recycled materials.

Different plastic composite materials, as well as unreinforced plastics, can be used to make up laminates. Glass reinforced materials can be combined with foamed plastics or even metal and wood (for example

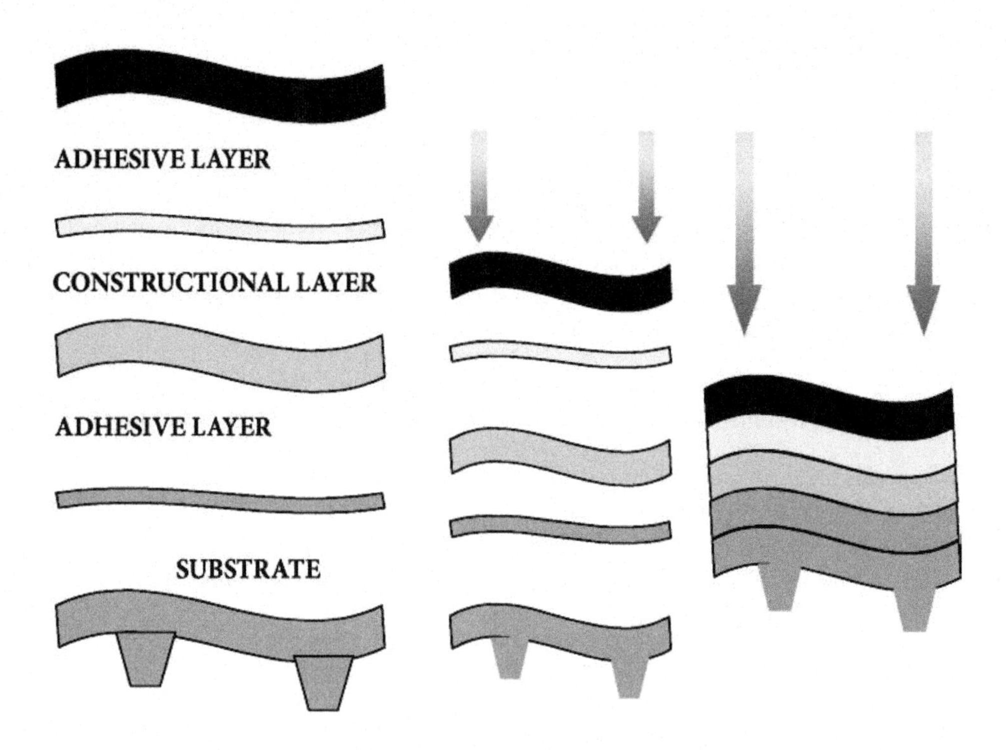

Figure 1.37

Typical laminate structure

in kitchen worktops and cladding panels). The behaviour and range of composites is a complex subject beyond the scope of this book and the interested reader should refer to more specialised publications for further discussion on this area.

1.7.2 Nanocomposites

Nano-sized particulate materials (a nanometre is 10^{-9} metre) are increasingly being used in plastics to impart improved mechanical or thermal properties to the base resin without increasing material density or reducing light transmission. By adding just 5% of a nanomaterial into a clear resin, the transparency is retained but the material has significantly enhanced properties.

In plastics, these nanomaterials have tended to be in the form of minerals such as clay. Using nanoclays with nylon 6 and PP in the automotive industry has enabled these materials to go under the bonnet where high temperatures had previously prevented their use. In packaging, nanoclays have been used to enhance stiffness and prevent ingress of oxygen (improved barrier properties) again in nylon 6. Other materials such as PET have also been investigated.

The potential advantages of using nanotechnology are highly interesting to scientists. The following headline from the ScienceDaily in 2007 gives a clue as to why this is.

> *ScienceDaily (Oct. 5, 2007) — By mimicking a brick-and-mortar molecular structure found in seashells, University of Michigan researchers created a composite plastic that's as strong as steel but lighter and transparent. (www.sciencedaily.com)*

While this research is in its early stages a plastic as strong as steel but light and transparent is very appealing. Imagine lightweight, fuel-saving, transparent cars or see-through bridges. Who knows what materials currently in development may enable future generations to achieve!

A further nanocrystalline metal/polymer hybrid material has recently been introduced onto the market by the company DuPont. This material is called Metafuse™. It combines light weight with the strength and stiffness of metal and can be shaped like a conventional thermoplastic. This is followed by an after-forming process that deposits a thin high strength metal layer creating a nano-laminate type structure.

Hybrids of complex nano-scale design combining materials such as polymers, ceramics and metals are likely to be of increasing interest to engineers in the future.

1.8 Thermoplastic fibres

Figure 1.38 shows various brushes which all contain plastic in fibre form (as do synthetic fabrics). Some thermoplastics when molten can be drawn into thin strands. Once cooled these are permanent (unless re-melted again). PET, PP and nylon 66 are examples, and roughly 15% of all thermoplastic production goes into the production of fibres. For plastics to be capable of producing fibres, they must have a crystalline structure. Therefore, fibres are produced from semi-crystalline plastics and not amorphous ones. As the material is stretched, (a process called drawing which will be mentioned again in Part 2) the material becomes uniformly thinner. In an undrawn material, the crystallites are likely to be randomly orientated with respect to one another, however as drawing proceeds more chains lie in the direction of drawing, and greater alignment occurs. This in turn increases the stiffness of the strand, and highly orientated materials provide incredible stiffness and strength as discussed previously in the case of Kevlar.

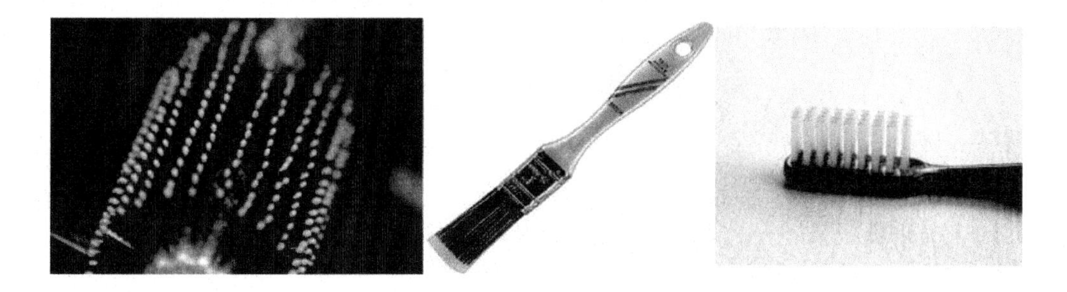

Figure 1.38

Various brushes with plastic fibre 'bristles': hairbrush (left), paintbrush (centre), toothbrush (right)

The properties of the drawn fibre depend on the original average chain length and also how far it has been stretched. This quantity is called the draw ratio and is defined as the final length of a fibre sample divided by its original length. Common draw ratios are between four and six. Typically, the fibres used to produce soft flexible weave for PET clothing will have a lower average chain length and a lower draw ratio than fibres used in engineering applications, which require higher stiffness and greater orientated crystallinity.

1.9 Biopolymers

Up until now the discussion of plastic and polymers has been based entirely upon synthetic materials which are created from fossil fuel feedstock. Whilst plastic production probably accounts for around 4% of all the oil that is refined, there is a growing trend to move away from using oil reserves in all aspects of modern living, including plastics, as well as transport and energy. To achieve a more sustainable solution to creating plastics, a more renewable polymer feedstock is required. Since the natural world is full of biopolymers, a number of likely alternatives have been suggested.

Biomaterials have been used almost as long as man's history has been recorded, amber is a natural resin, shellac has a long history and gutta percha (like natural rubber) is an extract from tropical trees. Bones, horns and gelatin can all be used as biomaterials.

A well known and long established commercial bioplastic is cellophane, a sheet material made from cellulose. The process usually begins by treating shredded wood pulp with caustic soda, although it can also be made from cotton. It has several very useful properties as it is transparent and glossy, and its stiffness allows it to stand upright. The material we see commercially often contains several layers (it is a laminate: see section 1.7.1.5) to impart properties such as resistance to gas permeation. A typical four-layer system could contain a wax moisture-proofing layer, a plasticiser, and a blending sheet. Alternative barrier layers can be used and the properties are comparable to high density polyethylene. Whilst this material is still widely used, its use in packaging has been reduced in favour of materials such as polypropylene.

One of the big selling points for modern bioplastics over synthetic plastics is their biodegradability and their perceived ability to disappear in landfill sites. Plastic waste gets a lot of bad press and for good reason. It is highly visible in both the waste stream and the environment. However, in the midst of all the political arguments raging on the future of plastics it should be remembered that plastics in the environment are not naturally occurring. It is we as consumers who are failing to keep our environment free from waste (and this is not confined to just plastic).

1.9.1 Plastic degradation

It is important in the context of comparing polymers and biopolymers to consider plastic degradation. Degradation can be considered as a harmful change in appearance, physical properties or chemical structure and can also occur in synthetic plastics as well as the so called 'green' plastics. There are a number of definitions as regards degradation, which will now be broadly introduced. There are a variety of international standards with which materials must comply to be certified as 'biodegradable' or 'compostable' for instance.

In **non-degradable** plastics, environmental stability is the issue and for manufacturers, the aim is to increase degradation not reduce it. They materials are water resistant and strong, and microorganisms do not attack them. These materials may remain in the environment in a recognisable form for large timescales (>20 years) if they are not recycled or reused. The majority of plastics fall into this category.

1.9.1.1 Readily degradable plastics

Readily degradable materials have all the properties they require for service, but after they have served their useful life they fall apart and hence return to the environment. Existing synthetic materials from fossil fuels, such as poly(vinyl alcohol), poly(glycolic acid) and polycaprolactone may serve this purpose. Generally these are speciality polymers rather than commodity materials and they may degrade in timescales of days or weeks. These are not biopolymers but synthetic degradable materials.

1.9.1.2 Controlled degradation

With controlled degradation, plastics are 'programmed' to degrade under a set of specific conditions after service has been achieved. One strategy is to use exposure to sunlight in order to trigger photodegradation. This is done by incorporating photosensitive components which encourage a process called chain scission. This breaks the polymer chains into smaller and smaller segments. Eventually a powder remains, which is dispersed into the environment by the action of wind and rain.

Commodity plastics such as polyethylene, polypropylene and polystyrene can be modified in this way. Additives can also be incorporated to trigger other degradation methods such as oxidation or hydrolysis. The main problem with these mechanisms is that you cannot control the environment and the weather! Accuracy in degradation time is therefore quite hard to achieve in the real world as opposed to that achieved in a controlled laboratory environment.

1.9.1.3 Biodegradation and biopolymers

This is a chemical degradation of the material caused by naturally occurring microorganisms such as algae, bacteria and fungi to produce carbon dioxide and/or methane. This depends on factors such as temperature, moisture and aeration. Biopolymers are inherently biodegradable as unlike synthetic polymers with carbon-carbon backbones, they also have oxygen and nitrogen in their backbones which is responsible for the biodegradability. Composting is a form of biodegradation that should leave no distinguishable residues or toxic remnants.

Biological polymers (biopolymers for short) are made from microorganisms, plants and animals and can be split into three general types; carbohydrates, proteins and polyesters.

Carbohydrate is by far the most abundant organic material type. This includes the polysaccharides which make up 75% of all organic matter, with the best known of these materials being cellulose from plants. A further non-carbohydrate component of plants is lignin which makes up 20%. Starch is also a carbohydrate.

Proteins are polymers formed by the polymerisation of amino acids, there are twenty amino acids which are found in all living organisms. Examples of proteins are collagen, casein and keratin.

Polyesters can be produced by bacteria and these can be produced commercially through fermentation in environmentally controlled bioreactors. These natural polyesters are biocompatible and biodegradable and so have medical applications. An example of these types of materials is polyhydroxyalkanoate, known as PHA. Types of PHA include polyhydroxybutyrate (PHB) and polyhydroxyvalerate (PHV). Further discussion is beyond the scope of this book.

For recycling of plastic consumer waste, the inclusion of readily degradable, controlled degradable and biodegradable materials can actually produce severe problems in controlling the quality of the plastic waste stream. This issue will be discussed further in Part 4.

1.9.1.4 Synthetic biopolymers

The terminology may seem confusing in this case, but there are some naturally occurring monomers that do not appear in nature as polymers. However, they can be polymerised in the laboratory. They have all the properties of biopolymers, and they are inherently biodegradable. Lactic acid is one example which is used to produce poly(lactic acid). Poly(lactic acid) known as PLA is a polyester like synthetically produced PBT and PET. A comparison of PLA and other packaging materials is provided in **Table 4.4.**

Like synthetic polymers, biopolymers can be used to make bioplastics by adding other additives.

1.10 Scope of plastic materials

Table 1.18 contains a list of abbreviations for some of the more common plastic materials. Less frequently seen materials are omitted. **Table 1.19** contains a list of commonly available copolymers. More about the applications of plastics in specific sectors such as automotive, construction, packaging etc can be found in Part 4. Some applications of common plastics can also be found in **Tables 1.10, 1.13, 1.14** and **1.15.**

Table 1.18

Abbreviations for common homopolymers, thermosets and chemically modified natural materials

Abbreviation	Polymer
CA	Cellulose acetate
CAB	Cellulose acetobutyrate
CAP	Cellulose acetopropionate
CMC	Carboxylmethyl cellulose
CN	Cellulose nitrate
CSF	Casein-formaldehyde
CTA	Cellulose triacetate
DAIP	Diallyl isophthalate
DAP	Diallyl phthalate
EC	Ethyl cellulose
EP	Epoxy
MC	Methyl cellulose
MF	Melamine-formaldehyde
PA	Polyamide (more often called nylon)
PA 6	Nylon 6
PA 66	Nylon 66
PA 11	Nylon 11
PA 12	Nylon 12
PAI	Poly(amide imide)
PAN	Poly(acrylonitrile)
PBT	Poly(butylene terephthalate)
PC	Polycarbonate
PE	Polyethylene
PE-C	Polyethylene (chlorinated)
PE-HD	Polyethylene - high density (more often seen as HDPE)
PE-LD	Polyethylene - low density (more often seen as LDPE)
PE-LLD	Polyethylene - linear low density (more often seen as LLDPE)
PE-MD	Polyethylene - medium density (more often seen as MDPE)
PE-UHMW	Polyethylene - ultra high molecular weight (more often seen as UHMWPE)
PE-VLD	Polyethylene - very low density (more often seen as VLDPE)
PEEK	Poly(etheretherketone)
PEI	Poly(etherimide)
PES	Poly(ethersulfone)
PET	Poly(ethylene terephthalate)
PF	Phenol-formaldehyde
PI	Polyimide
PIB	Poly(isobutylene)
PIR	Poly(isocyanurate)
PLA	Poly(lactic acid)
PMI	Poly(methacrylimide)

Table 1.18 cont'd ...

Abbreviations for common homopolymers, thermosets and chemically modified natural materials

Abbreviation	Polymer
PMMA	Poly(methyl methacrylate)
PMP	Poly(4-methylpent-1-ene)
POM	Polyacetal (also called polyoxymethylene)
PP	Polypropylene
PP-C	Polypropylene (chlorinated)
PPE	Poly(phenylene ether)
PPS	Poly(phenylene sulfide)
PS	Polystyrene
PSU	Polysulfone
PTFE	Poly(tetrafluoroethylene)
PU	Polyurethane
PVAC	Poly(vinyl acetate)
PVAL	Poly(vinyl alcohol), also seen as (PVOH) and (PVA)
PVB	Poly(vinyl butyral)
PVC	Poly(vinyl chloride)
PVC-C	Poly(vinyl chloride) (chlorinated)
PVC-P	Plasticised PVC
PVC-U	Unplasticised PVC
PVDF	Poly(vinylidenefluoride)
PVF	Poly(vinyl fluoride)
SI	Silicone
SP	Saturated polyester
UF	Urea-formaldehyde
UP	Unsaturated polyester

Table 1.19

Abbreviations for common copolymers

Symbol	Constituents
ABA	Acrylonitrile/butadiene/acrylate
ABS	Acrylonitrile/butadiene/styrene
AMMA	Acrylonitrile/methyl methacrylate
ASA	Acrylonitrile/styrene/acrylic ester
EVA	Ethylene/vinyl acetate
EVAL	Ethylene/vinyl alcohol
ETFE	Ethylene/tetrafluoroethylene
FEP	Perfluoroethylene/propylene
SAN	Styrene/acrylonitrile
SB	Styrene/butadiene rubber (more often seen as SBR)
SMA	Styrene/maleic anhydride
VCE	Vinyl chloride/ethylene

Part 2. Processing

2.1 The principles of plastics processing

Turning plastic materials into finished goods is a big global business.

In 2001, the production of 'miscellaneous plastics products' was the fourth largest manufacturing industry in the United States. In Europe in 2007, plastic converters (those making plastic feedstock into finished and semi-finished goods) comprised around 50,000 companies employing 1.6 million people. The turnover created by this activity was 280 billion Euros. In the United Kingdom 2.1% of the GDP is created by the plastics industry. This industry supports the employment of 220,000 people.

The continued success of the plastics industry relies on its ability to produce satisfactory products at a competitive price. For this strategy to be successful, it is necessary that the parts are of an overall suitable and consistent quality, the waste and rejects are kept to a minimum and that the parts have acceptable final properties.

A number of different processes are used, and the choice is dependent on a number of factors including the size and shape of the part required, the number of parts required, and the material of manufacture. **Table 2.1** gives an overview of eight of the most common high volume production processes with examples of common application types. Further details on these specific processes and others can be found in the relevant sections that follow. A continuous process here is defined as where the final end product is continually produced until production is stopped or raw material is exhausted. Extrusion is as continuous process. A semi-continuous process is one where the final components are produced intermittently during the process. Injection moulding is an example of this.

Before these processes are discussed further it is necessary to consider some basic properties of plastics that can be applied to all processing routes. All machinery for processing of plastics must be designed with these factors in mind.

These properties are:

- Low thermal diffusivity
- High viscosity
- Viscoelastic behaviour

The thermal diffusivity of polymers varies but is low. In polyethylene for example it is 1.3×10^{-7} m^2/s.

Table 2.1
Comparison of plastic shaping processes

Process	Pressure requirement (1-3, 3= highest)	Type of plastic [*Thermoplastic (P), Thermoset (S)*]	Foamed products possible	Example of products	Continuous (C) or semi-continuous process (S)	Rods, profiles, tubes, sheets	Complex geometry possible	Hollow containers
Extrusion	2	P	YES	Pellets, pipes	C	YES	NO	NO
Injection moulding	3	P and S	YES	Lids, caps	S	NO	YES	NO
Blow moulding	2	P	YES	Milk bottles	S	NO	NO	YES
Film blowing	2	P	NO	Carrier bags	C	NO: thin sheets	NO	NO
Thermoforming	1	P	YES	Plastic cups	S	NO	NO	NO
Compression moulding	2	P and S	YES	Trays	S	NO	NO	NO
Injection compression	2	P and S	YES	Trays	S	NO	NO	NO
Rotational moulding	1	P and S	NO	Water tanks	S	NO	NO	YES

When compared to a metal such as copper (1.13 x 10^{-4} m^2/s), the metal's value is a thousand times greater. In practical terms this means that metals will conduct heat quite quickly – making them ideal for saucepans for instance. If however you chose to make a plastic saucepan which was 1 cm thick, it would take well over 15 minutes for any heat to diffuse through to its contents. Even at a thickness of 1 mm this would still take 10 seconds. Therefore, when we consider ways to create molten plastic, relying upon conduction in containers would be expensive and time consuming, leading to long process times and low flow rates. Consequently, most processes which are used to melt plastics use equipment designed to treat the plastic in a thin layer – reducing the problem of thermal diffusivity and allowing all the plastic to attain the same temperature quickly.

The main heat transfer mechanisms employed in polymer processing operations are conduction (heating and cooling), convection (polymer and interface such as air or water) and viscous heating.

The high viscosity of molten polymers, (10^5 to 10^7 times greater than water), means that they have a higher resistance to flow than water, which necessitates high pressures to induce flow. If visualising the flow compared to that of water running down a very shallow slope, the polymer material would have a consistency varying from thick honey to cold modelling clay. More pressure is required than just gravity (unlike water), to get the material to flow.

During polymer processing operations, this creates pressure gradients: areas of high pressure and areas of low pressure. Applied pressure also causes the polymer chains to deform. The amount of pressure required varies according to the processing route. Injection moulding for instance requires very high pressures to produce components, blow moulding uses lower pressures, and in thermoforming the pressures exerted on the materials are very low. The nature and direction of this pressure can also vary between processes.

Consider a molten polymer material flowing though a narrow channel for instance. (Again a comparison with water in a pipe may be useful). The polymer flowing along near to the cavity walls is subject to more pressure than the material in the centre. This is because there is a resistance to flow induced by the side wall of the pipe. This type of pressure effect is called shear and the result of shear on polymers is to reduce the viscosity, as polymers exhibit behaviour called shear thinning. Viscous heating is also generated as the polymer flows. Hence, viscous heating leads to differential temperature gradients across channel flow. Since the thermal diffusivity of polymers is low, it takes time for temperatures to equalise across a channel such as this.

As polymer melts experience shear thinning, the actual shear rate present is an important factor during processing, and the viscosity of the material used (at the relevant temperature) becomes an important material property for a designer to consider. Very high shears are experienced by material during injection moulding processes. Material shear is also experienced during all extrusion and moulding operations but to a lesser degree.

Polymers behave in a viscoelastic manner, that is they show both viscous properties (they have a high resistance to flow) and elastic properties (they are able to return to their original shape after deformation such as compression or stretching).

The viscoelastic properties of polymers can create a number of flow phenomena during processing. This very much depends on the direction in which the polymer is exposed to applied pressure. This can be exerted in a number of directions: uniaxial, biaxial or shear as shown in **Figure 2.1**. Each results in certain responses from the polymer.

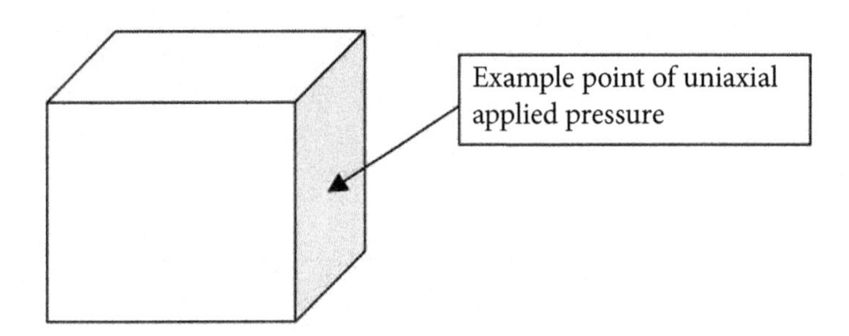

Pressure can be exerted on or across any of the six faces of a cube. If exerted across just one axis it is a uniaxial pressure, across two it is a biaxial pressure.

If one face is fixed but the other is moving, shear flow (and deformation) are produced.

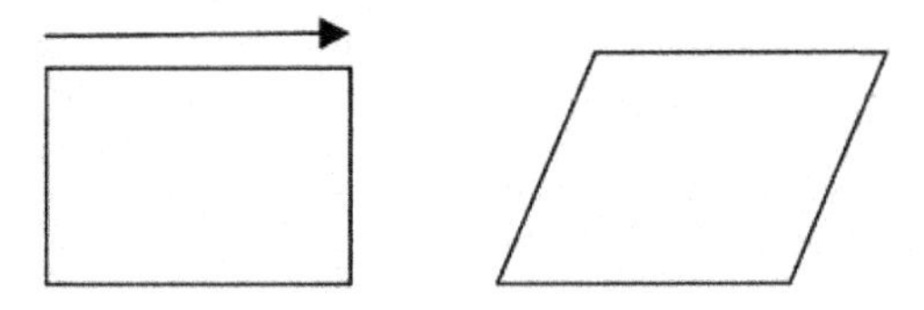

Figure 2.1

Directions of applied pressure (uniaxial, biaxial or shear)

For example inflation processes (see film blowing, blow moulding) exert biaxial pressure in an action similar to blowing up a balloon (**Figure 2.2**). Biaxial pressure causes extensional flows with a resultant thinning of the wall thickness of the component. This is also sometimes referred to as elongational deformation.

In shear flow, as experienced by polymer moving along a channel, the stress is exerted across the surface of the material, leading to deformation. Shear flows are experienced in different degrees in common processes such as extrusion, injection moulding and compression moulding.

For thermoplastics and thermosets there are distinct differences in processing routes due to molecular differences and their different heating requirements. Thermoplastics need to be heated up to a point where they can be forced to flow under pressure into a cooler tool cavity or air. Once they are formed, they need to be cooled until they regain enough strength and stability to retain their intended shape. With thermosets, the crosslinking reaction requires more heat than that required during the flow

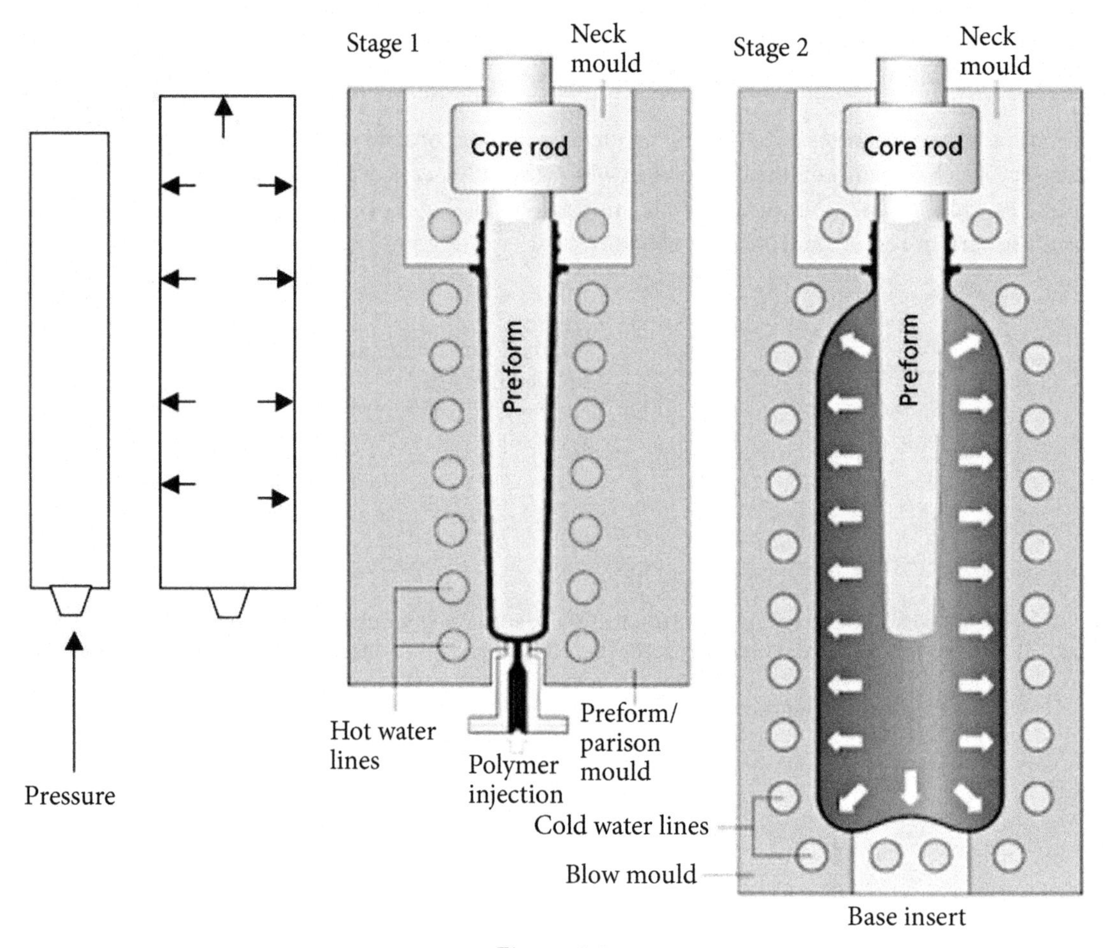

Figure 2.2

Biaxial deformation (illustrated by injection blow moulding)

process. Therefore, thermosets go into heated tools. These different heat requirements will be discussed in the relevant processing sections.

There is some overlap as some processes such as injection moulding and compression moulding are applicable to both sets of materials, although the manufacturing process itself is slightly different for the reasons just outlined.

2.1.1 Extrusion

The most common of all plastic processing routes is extrusion. This is because extrusion is used to make the plastic raw materials for other production processes such as injection moulding, blow moulding and film blowing. Extrusion is also used to produce a number of finished or intermediate plastic products such as plastic profiles, pipes, board and sheet (which is also then used as the feedstock for thermoforming).

Extrusion can be used to mechanically melt, mix and solidify polymer and additives together, for example to add a colour or colours to the final plastic, or to add reinforcement such as glass fibre. There are many different types of formulations (recipes!) for commercial plastics which can contain numerous components. These are all combined during the process of extrusion.

An extruder must perform the task of conveying, melting and mixing the raw materials to form a consistent (homogenised) product. The material must then exit the extruder where it can be shaped, cooled and solidified. Therefore, a basic schematic of an extrusion process can be represented as shown in **Figure 2.3**.

When considering a basic extruder as shown in **Figure 2.4** it can be seen that these areas relate to the hopper, extruder throat, screw (melting and mixing) and die (shaping) respectively.

An extruder can be one of two distinct types: a single screw or a twin screw, and the distinction will be highlighted later. The following descriptions of extruder components focus on the simpler single screw machine. However it is first necessary to consider the material preparation stages, drying and mixing.

2.1.1.1 Drying

Before material is fed into an extruder, it may need to be prepared. This can include drying to remove excessive moisture, or preparing a mixture of materials to extrude. A number of plastic materials

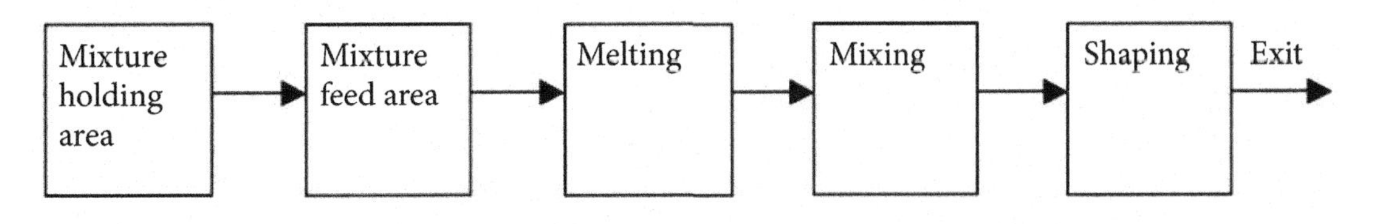

Figure 2.3

Basic extruder functions

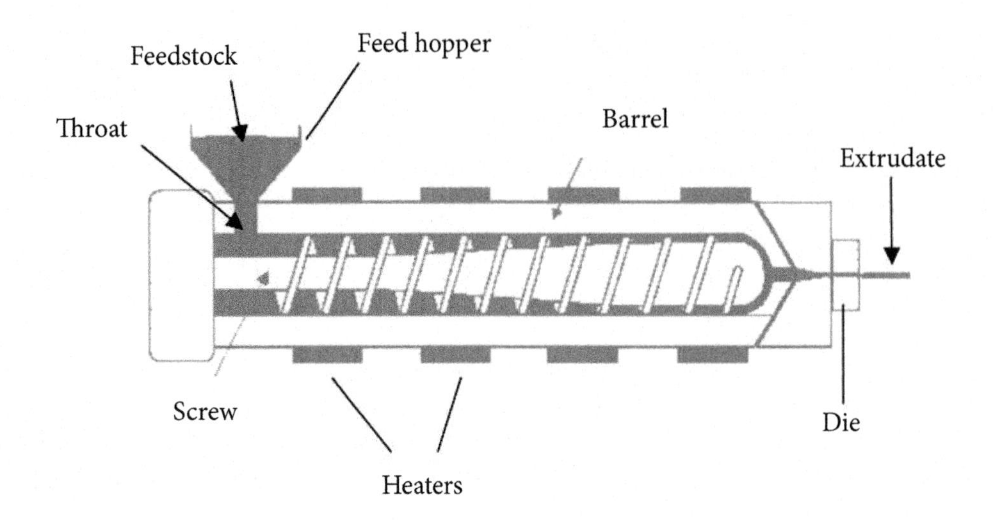

Figure 2.4
Single screw extruder

are hygroscopic – that is they absorb water, and this can cause problems with both processing and/or the final part. Examples of hygroscopic material are nylon and polycarbonate. Here processing a wet material can cause splash marks to be visible in the final component. In the case of PET or PBT, using material containing moisture when formed can seriously affect the mechanical properties of the final component. Pre-drying is therefore essential and can be carried out using dehumidifying dryers. Other materials may also require drying if they have picked up excess moisture over time, for example if bags of plastic have been stored outside a factory for long periods.

2.1.1.2 Mixing

Uncoloured plastics may be pigmented a huge variety of colour shades by using a colour masterbatch (concentrated colourant predispersed in a plastic). This would need to be added at a set percentage in the mix to achieve the required colour, before it is loaded into the extruder. Alternatively, a manufacturer may want to add another additive or some recycled material to the mixture. There are two ways materials can be prepared; either by cold mixing or hot mixing. Cold mixing is simply a blending of components at room temperature. For small batches this can be done by simply putting the materials in a bag and manually shaking them together, however for large, heavy mixes automated machinery such as a free-falling mixer (tumble mixer) is used. In this process the mixing cavity is simply rotated by a motor and gravity does the mixing.

Large mixing devices can be placed above the extruder hopper to allow direct transfer of the mixture to the extruder or metering device. Hot mixing can also be carried out this way, where the mixing is done at a temperature which will melt the components together, and this can also be fed directly to an extruder under gravity.

2.1.1.3 Hoppers – mixture holding

The extrusion process starts at the hopper, and all extruders need hoppers. These hold the plastics prior to extrusion. The feedstock in the hopper must be able to pass through into the extruder and is therefore sized accordingly in the form of pellets, powder, or molten polymer. A hopper is loaded either by hand or by pneumatic conveying devices and extrusion can then commence. Simple hoppers work by gravity feed and material falls into the next stage – the extruder throat. As well as a material exit point at the bottom of the hopper leading to the extruder throat, hoppers also have a second material exit. This is to allow operators to empty them again. This is in case the wrong materials are loaded or excess material remains in the hopper when a job is completed.

2.1.1.4 Extruder throat – mixture feeding

Below the hopper, there is a simple hand push or automatic slide device which, when open, allows material to drop into the extruder throat. This lies between the hopper and the extruder screw beneath. This area can be quite critical for some materials to enable extrusion to be carried out successfully. The reason for this is that whilst the hopper is at room temperature, the extruder screw beneath it is heated. This temperature can vary depending on the plastic material being extruded but as a guide 180 °C – 300 °C covers most of the more common plastic materials encountered. The extruder throat, sitting between room temperature and high temperatures, is subject to heating and if there is no control of temperature this can cause material to melt prematurely and block the throat. It is therefore necessary to keep this area slightly cooler and the temperature control used for extrusion is therefore often staggered with cooler temperatures at the back of the extruder screw.

2.1.1.5 Extruder screw – melting and mixing

From the extruder throat the material drops onto the extruder screw. In a single screw extruder there is just one, in a twin-screw there are two. Twin screws are described in more detail in the following section. In a single screw machine, an Archimedean screw design is used to convey the material along while melting and mixing. It is contained within an extruder barrel and heating zones in the form of heater bands are screwed onto the extruder barrel to provide heat, as shown in **Figure 2.4**.

The speed of the screw can be varied and is usually measured in revolutions per minute (rpm). This is displayed to operators for adjustment along with temperature controls and also the force (torque) on the screw system which is created by a combination of molten polymer, temperature and screw speed. If the maximum safe load on the system is exceeded by this combination, then the machine trips, shutting down the screw to prevent it from breaking. The critical time for torque is during start up of the machine. Once running, unless there is a major technical problem or operator error (such as a metallic object accidentally finding its way onto the screw for instance), torque levels should be steady.

The temperature of the screw is measured by the series of thermocouples attached to the barrel. The number of these will vary depending on the size of the machine but they will generally correspond to any changes in the function of the screw, such as material metering, plastication areas[a] and exit points.

---

a. This is where the material is plasticised. A material with plastic properties is able to be shaped and formed

During extrusion temperature is monitored by control devices that switch between turning on the electrical heating zones and turning on screw cooling. Water cooling is used to remove excess heat. Cooling systems on extruders are essential, without cooling the action of mixing and the viscosity of the materials would cause the material to overheat. The heating is used to preheat the machine prior to starting extrusion. Any attempt to convey an unmelted polymer in an extruder would simply trip the safety units of the machine. Once extrusion commences, temperature control tends to be for cooling not heating.

Screw designs vary depending on the plastic material being extruded. This is because it is possible to vary the function of the screw in different places along its length. There can be feeding areas, melting areas, compression areas and mixing areas all combined on one screw (**Figure 2.5**). The length and depth of these channels depend on the materials used, the amount of mixing, and the output rates required. Therefore, screw configurations are highly variable in design.

It should also be remembered that polymers are not good heat conductors and the polymer thickness in any section of the screw is kept very low by design. The clearance between the extruder screw and the extruder barrel wall is therefore also considered when designing machines.

Three terms are often used in relation to screws: passive, standard and barrier.

A passive screw is used simply to melt and transfer material (also called a metering screw). A standard screw will contain a number of regions with differing configurations which carry out tasks such as mixing and produce compression areas as shown in **Figure 2.5**. A barrier screw configuration contains an additional flight which separates molten and solid materials into different channels. The geometry of these channels ensures complete melting and improved mixing.

A further common term relating to extruder screws is the L/D ratio. This refers to the ratio of its length (L) to its diameter (D) e.g. 24:1, 20:1.

2.1.2 Twin screw machines

The fundamental operation of a twin screw machine is the same as a single screw extruder, however instead of one single screw within the barrel, there are two. Twin screw machines operate in two distinct

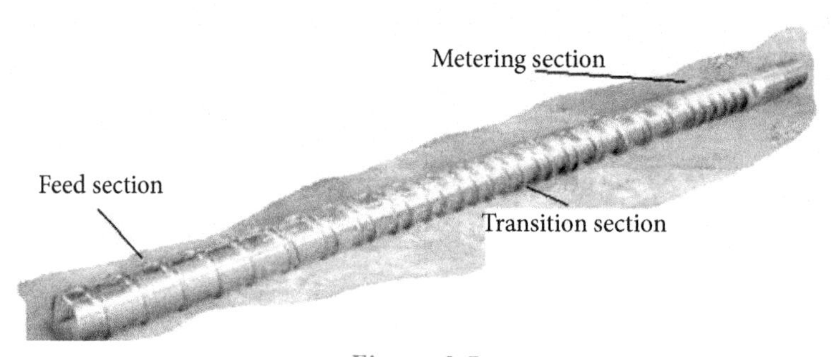

Figure 2.5

A screw with three areas: feed, transition and metering

ways; co-rotating or counter-rotating. Each has different usage. Co-rotating machines improve mixing and are used in the production of compounds, whereas counter-rotating screws are used for heat sensitive materials such as PVC (**Figure 2.6**).

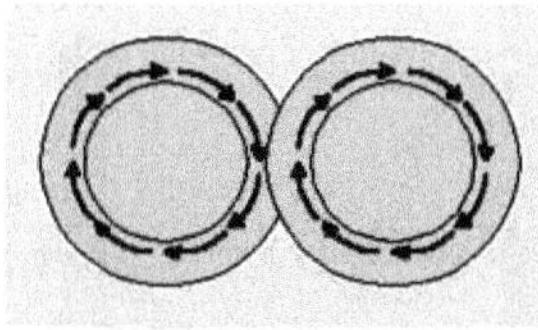

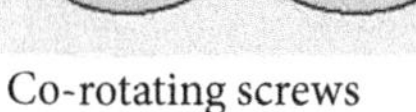

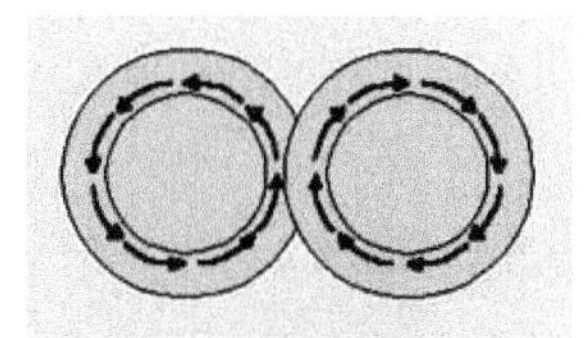

Co-rotating screws Counter-rotating screws

Figure 2.6

Directions of extruder screw rotation

Because of the way these machines work it is not possible to simply drop material into the feed throat by gravity as in single screw extrusion. It is necessary to dose the amount of material that is fed into these machines, and therefore devices such as gravimetric or volumetric feeders are used to dose the materials falling onto the screw. A further advantage of twin screw extrusion is the ability to feed separate components onto the screw using two separate feeders. Glass fibre filled compounds are produced in this way with the polymer and other additives fed through one hopper, and the glass fibre then fed in separately downstream.

Heating zones operate in a similar manner to those described in the single screws. In most other ways, the actions and operation are very similar, in conveying, melting and mixing. One area in which they may vary is the general usage of vent ports. These are areas of the screw that are open to air to allow volatiles (water vapour and other gases) to escape the processing operation. Some gases may be drawn off rather than vented to air. Vent ports are more common in twin screw extrusion than single. This is due to the greater material complexity in twin screw formulations.

2.1.3 Processing beyond the screw

2.1.3.1 Breaker plates

Once past the screw, twin screw and single screw extruders have the same components. The rotation of the screw or screws causes rotation of the polymer. To make the direction of flow more perpendicular towards the exit, the flow must be redirected and the rotation of the melt removed. This is the purpose of the breaker plate. If there was no breaker plate, the polymer would continue to spiral as it exited the machine in a similar manner to the action of the screw. A breaker plate can also act as a flow restrictor, allowing melt pressure to build up behind the plate, and this is also where screen packs can be used. As a breaker plate appears effectively like a sieve (**Figure 2.7**), screen packs can provide even finer sieving. These mesh grills also encourage melt pressure build up and filter out impurities and agglomerates.

2.1.3.2 Die (forming and exit)

The die is the exit point from the extruder and where the plastic is shaped to form the desired end product. The die acts as a flow restrictor and pressure is built up behind it, this is called die head pressure. When the melt emerges, its shape must be fixed fairly rapidly and it will have a constant

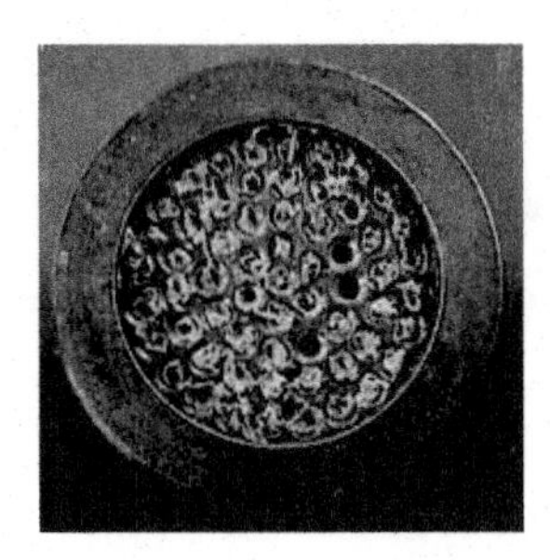

Figure 2.7

Breaker plate (with solidified polypropylene blocking some holes)

cross section based on the die shape regardless of whether it is a sheet, pipe or rod.

For example if material is being compounded, a strand die may be used to form thin strands which can then be cut externally into small pellets. If hollow profiles (e.g. window frames) are being made, the die will have this shape. Pipes, film, sheets and filaments can also all be formed by extrusion through a correctly designed die and suitable downstream equipment. A pipe extrusion die is shown in **Figure 2.8**.

2.1.3.3 Cooling and external cutting devices

On exiting the die, the polymer needs to be cooled to solidify it. Eventually ambient air would do this job, but this is far less efficient than using water cooling. Therefore, the emerging material is fed into a cooling bath. As the process is continuous, the plastic can be either immersed in water before being fed into a cutting device or water can be sprayed onto the surface. In profile extrusion where continuous lengths are being produced, downstream equipment must match the extruder output. The specific cutting devices used will depend on the end product and may include cooling tanks and sizers, cutters, pullers (to pull the extrudate along at the required regulated speed) or coilers (used to coil tubing onto reels). Profiles can be cut into long lengths controlled by machines called haul offs. Pellets can be produced using pelletising devices.

2.1.4 Compounding

When extrusion is used to produce pelletised raw materials for other processes (generally individual pellets which are pea-sized or smaller), this is referred to as making a compound. A compound 'recipe' may contain any number of additives as illustrated in Part 1. Often the level of residual polymer in the material will determine whether single screw or twin screw compounding is required. All formulae can generally be made on a twin screw machines but very heavily filled materials cannot generally be made on a single screw machine. By first pre-extruding all the components of a polymer-based compound together the mixture can be homogenised to ensure all components are fully dispersed. The screw acts as a mixer and evenly distributes the materials, which can have widely differing masses.

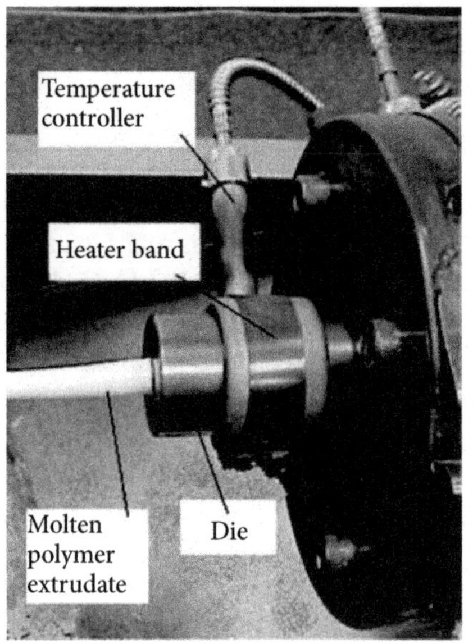

Figure 2.8

Pipe extrusion showing the die

When incorporating glass fibre, for example, this cannot be premixed with polymer before extrusion as the fibres act as bridges in the feed throat and block the extruder hoppers. In this case it is necessary to use two hoppers which are metered to ensure that a correctly proportioned mixture of the two materials enters the extruder.

- Hopper 1: Supplying 70% of total mixture, made up of 69% PP, 0.5% heat and light stabilisers and 0.5% colourant
- Hopper 2: Supplying 30% glass fibre

A twin screw machine is used and the metered material in Hopper 1 is fed as described previously. Downstream of this is Hopper 2, which feeds the glass fibre into the extruder at a second feed port. The glass fibre is then homogenised into the mixture before the homogenised compound exits at the die to be cut into pellets. A schematic is shown in **Figure 2.9**.

Pelletising is a matter of sizing. If pellets are too big they will not feed through the hoppers of machines such as injection moulders and blow moulding machines. Extruders also require a suitable sized feedstock and this is why recyclate materials need to be shredded to ensure that they too can be fed back into processing machinery. There are many commercial compounds which exist in the marketplace supplied by both resin suppliers and specialist compounding companies. Custom compounds can generally be supplied on demand.

2.1.5 Coextrusion

Coextrusion is an extrusion process in which two or more materials are extruded together to produce multiple layers (or similar effects) in one product. The materials remain distinct from each other, i.e. they do not mix, but as they contact when molten, a bonded interface is formed on cooling. This method is used either when all the requirements of a product cannot be met by a single material alone or for cost reduction purposes.

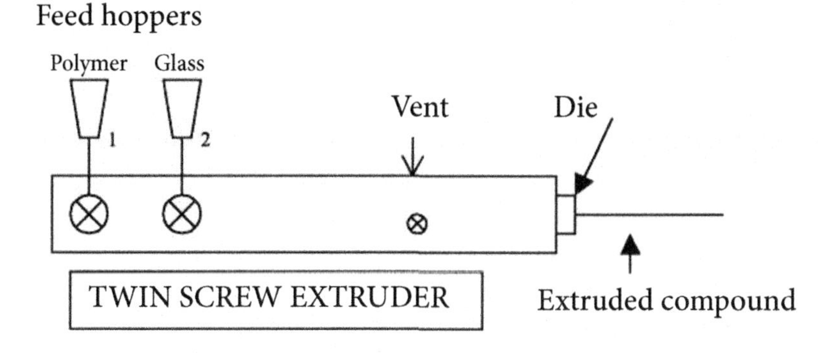

Figure 2.9

Two hopper compounding set up

Coextrusion can use single or twin screw extruders or a combination of both. Each of the materials needs its own extruder. The separate melt streams are then brought into contact in a component called a manifold die, which integrates the combining flow paths and the plastic exit point. The design of the manifold die controls the placement of the different materials and can be used to make stripes (as on a striped drinking straw) or layers (such as a multi-layer sheet) for example.

Imagine you wish to make a blue hollow pipe with a yellow stripe along its length – this can be coextruded. Each colour material has an individual extruder, and because the volume of yellow material used is less, the 'yellow' extruder would probably be much smaller than the blue one, and would be called the side extruder. Melt temperatures and flow rates are separately controlled until the two extruders are joined together, sharing a common die. It is here where the two materials meet and fuse together to form a finished part before exiting the machine. This is shown in **Figure 2.10**.

Profiles can also be coextruded using two different materials, resulting in properties that would not otherwise be possible. For example, PVC profiles can be produced with a foamed inner layer and an unfoamed outer layer to reduce weight and cost. Combinations of rigid and flexible PVC are also possible. Multi-layered sheets can be produced to utilise the properties of barrier layer materials such as EVAL (ethylene-vinyl alcohol copolymer) between HDPE layers.

Coextrusion manifolds can be used to produce a variety of multi-layer materials and products, indeed any shape created by extrusion can generally also be made by coextrusion. Other processes which start with an extrusion stage, such as film blowing and blow moulding, can also exploit coextrusion to manufacture multiple material structures.

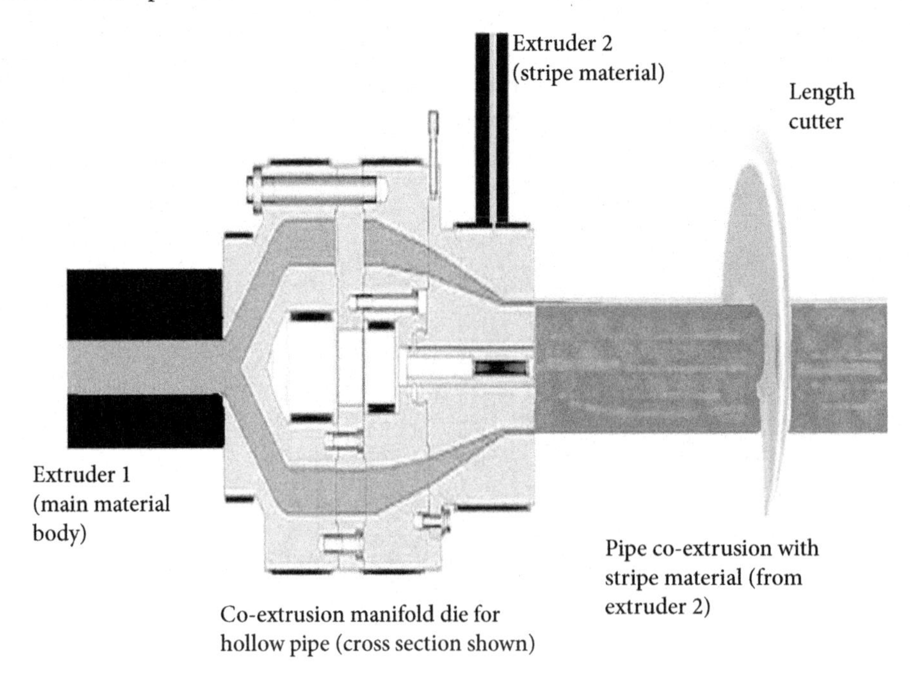

Figure 2.10

Coextrusion die

The discussion here has been focused on thermoplastic materials. This is because extrusion is not relevant to thermosets in this context. However extruders which work on similar principles to the ones described here are used to produce a number of thermoset raw materials such as powder coatings. These can be applied onto surfaces (such as a bicycle frames or cooker casings) and then fully cured by heat or UV light. It should be remembered that once cured thermosets cannot be remelted. Therefore, in producing powder coatings in an extruder it is essential that they are not cured. However further discussion of these production methods is beyond the scope of this book.

2.1.6 Injection moulding

The first injection moulding machine was patented in 1872 by John and Isaiah Hyatt. It has gone on to be the most important process after extrusion for the creation of plastic parts and components. Non-plastics as well as both thermosets and thermoplastics can be injection moulded. Elastomers, metal powders and ceramics can all be processed by this method, although of course only plastic moulding will be considered here.

Injection moulding is a very economical method of mass production. High volumes of completed parts can be produced rapidly directly from the machine, with very high precision tolerances from part to part. It is often not necessary to finish the components once they are moulded, however there are both in-process and after-process decoration options for manufacturers to consider if necessary. Decoration techniques are discussed in section 2.2.5.

Individual machines are configured for a certain type of material and size of component, and differ considerably in size, see **Figures 2.11** and **2.12**. There are machines capable of moulding at a micro and even nano scale. This allows injection moulding to be used to make tiny components which are even dwarfed by ants at shot weights of below 1 g. For example these may be tiny medical components which need to be as small and non-invasive as possible, or cogs and wheels for watches and electrical items. An example is shown in **Figure 2.13**. At the other extreme the biggest machines can produce components using up of 150 kg of material at a time with a clamping force of 55,000 kN. Such machines can produce giant underground tanks for example.

There are also a number of advanced plastic processing techniques based on injection moulding technology which enable multi-layered, hollow and foamed products to be made for example, and some of these will be discussed in a little more detail at the end of this section.

Whatever the size of the finished products, in order to produce injection mouldings the machine must perform three basic operations:

- It must plasticise and homogenise the feedstock.
- It must inject it into a mould tool at high pressures (500-1500 bar) to produce rapid mould filling.
- It must cool and eject the moulded component.

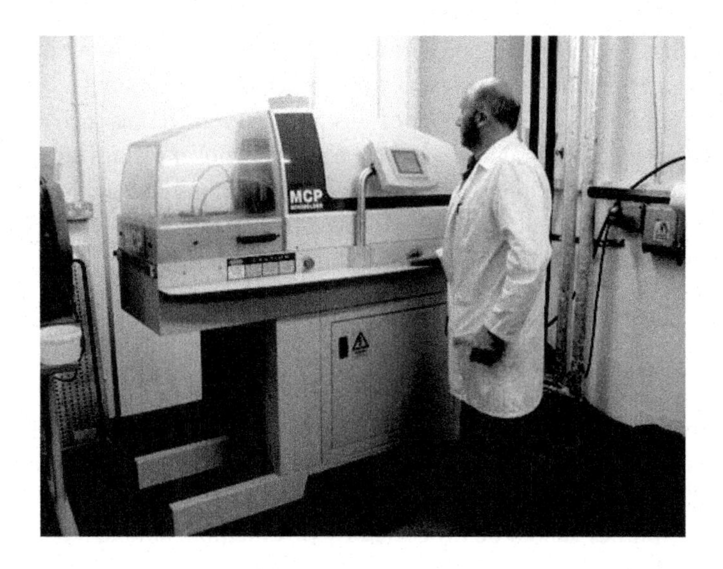

Figure 2.11

Micro moulding machine

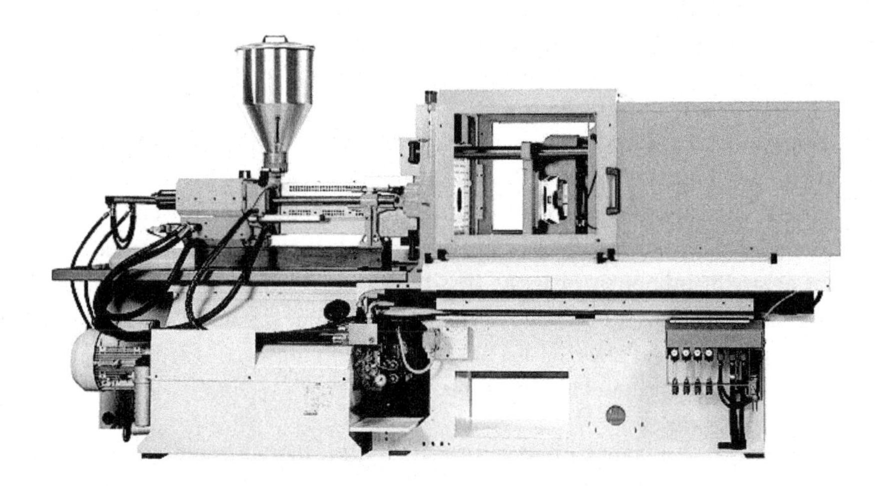

Figure 2.12

Commercial injection moulding machine
Image courtesy of ARBURG GmbH

There are intrinsic differences in the injection moulding needs for thermoset and thermoplastic materials, which require different types of machines. The biggest difference is in heating requirements:

- A thermoplastic material is heated in the extruder barrel to a temperature where it is able to flow under pressure. It is injected into a cooler forming tool, where it cools, solidifies and is ejected.

- A thermoset is kept molten but cool in the barrel to prevent premature curing and is injected into a hotter tool, where it crosslinks. Once the reaction is complete, the part is ejected.

Because of the differences in heating regimes, specialised machinery is used for each type of material. However the main components of each are similar.

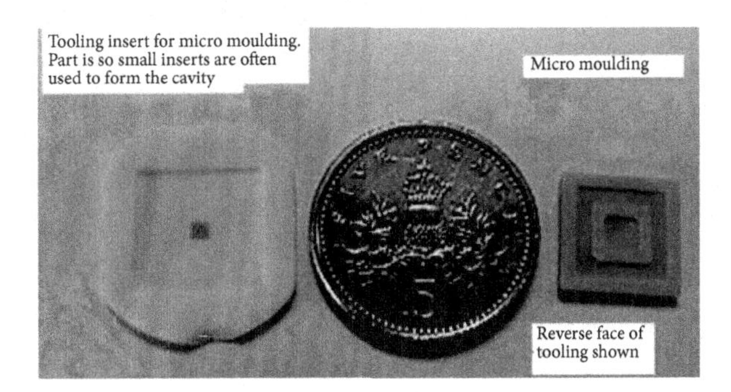

Figure 2.13

Micro moulded component

2.1.6.1 Machinery

Injection moulding machinery can be split into distinct units including the **injection unit** (**Figure 2.14**) and the **clamping unit** (**Figure 2.15**). Other components of the injection machine are the control cabinet and machine base. The control cabinet is where the operator sets and monitors control parameters, and the machine base holds the entire unit and can be driven by hydraulics (all older machines are driven this way) or electrical means.

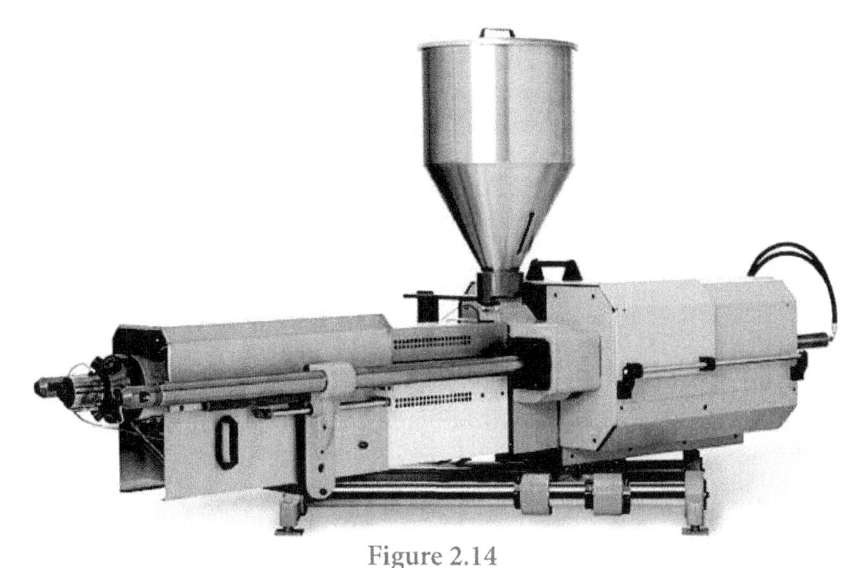

Figure 2.14

Injection unit

Image courtesy of ARBURG GmbH

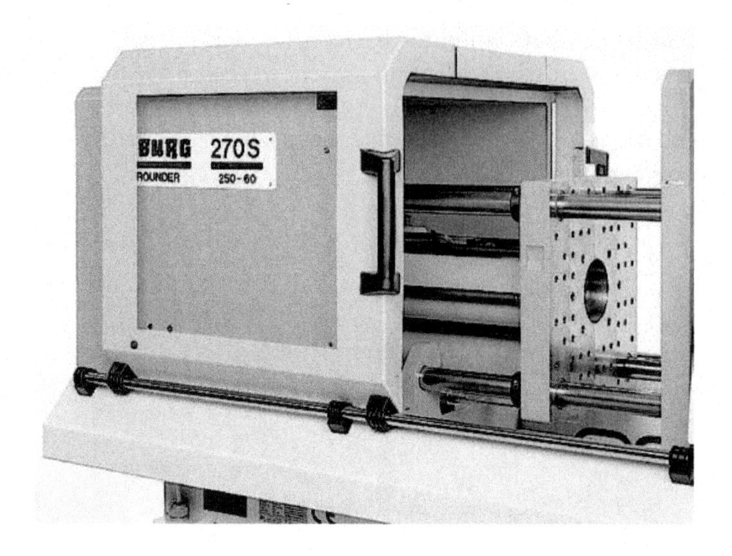

Figure 2.15

Clamping unit
Image courtesy of ARBURG GmbH

The job of the injection unit is very similar to that of a single screw extruder, in that it must feed, mix and convey material from one end of the screw to the other. However, the big difference is that the injection unit must also inject a controlled dose of molten material into the tool at high pressure, metered by the screw. This controlled dose is called the **shot weight**. Therefore, whereas extrusion is a continuous production process, injection moulding is not. Injection moulding screws which both plasticise and inject the material are called **reciprocating screws** (**Figure 2.16**). This is not the only type of screw used in injection units, but is by far the most common. Alternatively there are **plunger screws.** As the name suggests these simply force material into the cavity without mixing. Plunger screws are sometimes used in thermoset machines but rarely with thermoplastic materials. The screw is contained within an injection cylinder and this is where external heating units can be fitted to give controllable heat to the injection screw.

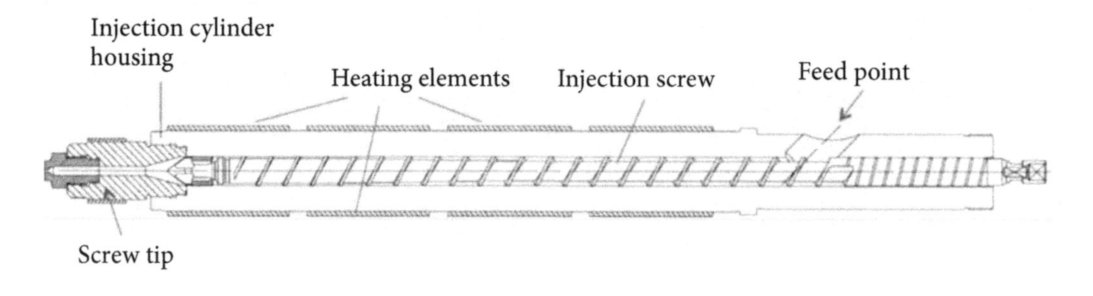

Figure 2.16

Reciprocating injection screw and cylinder
Diagram courtesy of ARBURG GmbH

The clamping unit is where the injection mould tool is sited. These are generally made of two halves which clamp together. One half of a mould tool for shoes is shown in **Figure 2.17**. The clamping unit is used to control the opening and closing of the mould tool. It is also used to control part ejection from within the mould tool. The clamping unit is sized so that it can exert enough **clamping force** on the mould tool to hold it closed during injection. Injection moulding is done at very high pressures and a large force is required to hold the mould tool closed; otherwise the pressure of the melt would simply push the two halves apart.

Figure 2.17

One half of an injection mould tool cavity for shoe soles

2.1.6.2 The injection moulding cycle

The action of producing one component from start to finish is called the **injection moulding cycle** and would typically consist of the following operations:

1. The mould tool closes.
2. A dosed shot of material is injected into the mould tool.
3. The material cools in the forming mould tool, during this time the injection unit must prepare the next dose of material.
4. The mould tool opens and the part is ejected.

By repeating this process over and over again, many parts can be produced in a rapid manner. An entire cycle may be carried out in less than 30 seconds so many components can be produced very quickly. The cycle time is dependent on the time taken for the moulded component to be ready for ejection. That can be after cooling to a suitable temperature where the plastic is solid (thermoplastic) or curing to a suitable level of rigidity (thermoset).

As a cycle begins and the mould tool closes, the injection unit is ready to inject. As plastics are so viscous, there must be enough initial force generated by the machine to overcome the resistance of the material to flow, therefore very high pressures are needed for this operation. This **injection pressure** varies according to the material, some values for common materials are given in **Table 2.2.**

The important parameters at this stage are initial barrel temperatures and the velocity of the molten polymer into the mould tool, which is called the **injection speed**. Speed is related to the viscosity (affected by temperature) and also shearing of the melt as the screw moves forward rapidly to deliver a controlled shot into the injection mould tool.

The material moves rapidly from the injection unit into the injection moulding tool in the clamping unit. In doing this, the material is moving from a region of high pressure (the injection unit) to a

Table 2.2

Control value ranges for injection moulding of some common materials

Material	Cylinder temperature (on injection barrel) (° C)	Mould temperature (°C)	Injection pressure (Bar)	Packing pressure (Bar)	Back pressure (Bar)
ABS	180-260	50-85	650-1550	350-900	40-80
LDPE	210-250	20-40	600-1350	300-800	40-80
PET	260-280	20-140	800-1500	500-1200	60-90
PP	220-290	20-60	800-1400	500-1000	60-90
PS	160-230	20-80	650-1550	350-900	40-80
PVC (flexible)	150-170	20-60	400-1550	300-600	40-80
PA 6	230-260	40-120	450-1550	350-1050	40-80
PA 66	270-295	20-120	450-1550	350-1050	40-80

region of low pressure (the mould tool). However once injection commences, the pressures begin to equalise. The clamping force exerted to hold the mould tool closed is needed at this stage or the force of injection would open the mould tool.

The movement and filling of the plastic into the mould tool occurs by a process called fountain flow. Mould tool filling can be seen in more detail in **Figures 2.18** and **2.19**. Due to the high pressures applied this happens very quickly, generally in under a second. A high-speed camera is required to capture the process. Material is pushed out to the walls of the mould tool as the newer and hotter material penetrates the centre of the melt stream.

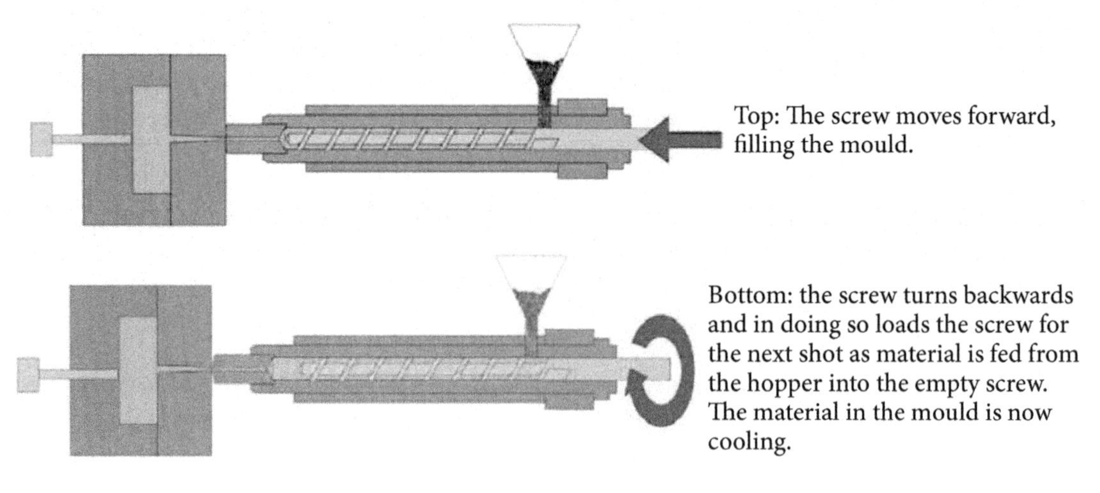

Figure 2.18

Injection mould tool filling and screwback
Diagrams courtesy of ARBURG GmbH

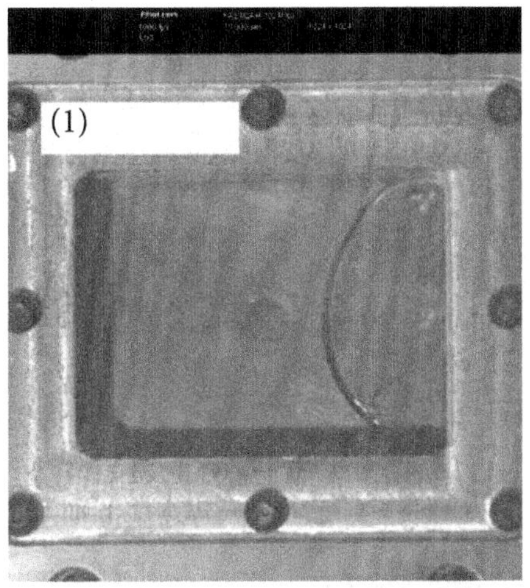

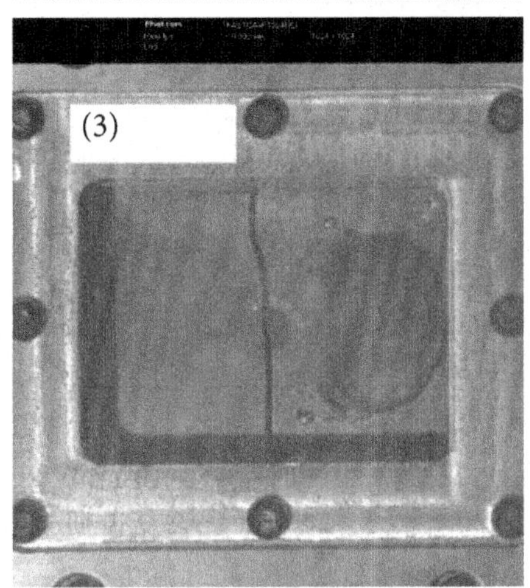

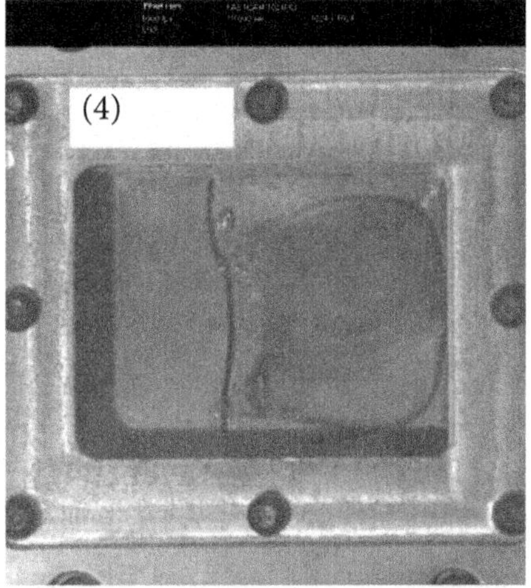

(1) At 0.011 seconds skin only injection has begun.

(2) At 0.232 seconds core can already be clearly seen (delay set was 0.1 seconds).

(3) At 0.262 seconds air bubbles can be seen.

(4) At 0.340 seconds an air bubble breaks the surface of the skin material.

(5) At 0.475 seconds mould filling is nearing completion.

Figure 2.19

Stages of mould tool filling (this example took less than 0.5 seconds)

Once injection has taken place the injection screw needs to screw back to prepare the next shot. During this screwing back process, further control on the quality of the melt can be employed. The time of screw return can be controlled and there are also mixing controls such as **back pressure** that can be employed. Back pressure can be controlled by the injection unit and as this level is raised a greater pressure is applied to resist the return of the screw. Therefore more turns of the screw are required to travel the same distance, with the effect that greater mixing of the melt occurs in the screw. This greater mixing can be beneficial for a number of reasons. One common example is when a colour concentrate called a masterbatch is used. By using a higher back pressure during screwback, a more consistent dispersion of the colour can be achieved in the final moulding.

2.1.6.3 Orientation, shrinkage and warpage

A further issue with injection moulding is one of orientation: because of the nature of filling, the plastic (both polymer and filler) will be orientated in the direction of flow. This can lead to dimensional shrinkage differences in line with and perpendicular to the flow. Good design can minimise problems related to differential shrinkage, but it cannot always be prevented from happening.

Different plastic materials have different levels of shrinkage as they solidify. Semi-crystalline materials shrink far more than amorphous materials due to molecular realignment and crystal formation. With injection moulding, these effects can result in sink marks on the surface of a moulding. To overcome this effect, immediately at the end of the high-speed injection, further material can be injected to compensate for the shrinkage that occurs as the material cools. This is always needed when materials are likely to shrink.

Therefore after the injection mould tool has initially been filled by high pressure injection, a second stage of injection called the **packing stage (or holding stage)** is initiated. The machine is preprogrammed by the operator to switch from a high-speed and pressure injection phase, to a lower speed and pressure packing phase. This forces a small amount of extra material into the mould tool to compensate for the shrinkage and ensure good packing of the mould tool is achieved. The **packing pressure** and the **packing time** can both be controlled, as can the point during the process at which that the machine makes the switch from the higher pressure to the lower.

Once the injection (and packing) stages have completed, the cycle enters the cooling stage. The plastic material begins to cool the moment it hits the injection moulding tool wall. However, before ejection it must be rigid enough to hold its final form without distorting (warping), i.e. the structure must effectively be frozen in. The temperatures for injection moulding ejection vary widely from material to material but the tool must not be opened until the appropriate temperature has been reached. Ejection is shown in **Figure 2.20**.

The injection mould tool temperature is controlled to optimise the cycle time and cooling time must be kept to a minimum to ensure a fast cycle time is achieved. The cooling stage accounts for the largest portion of the entire injection moulding cycle.

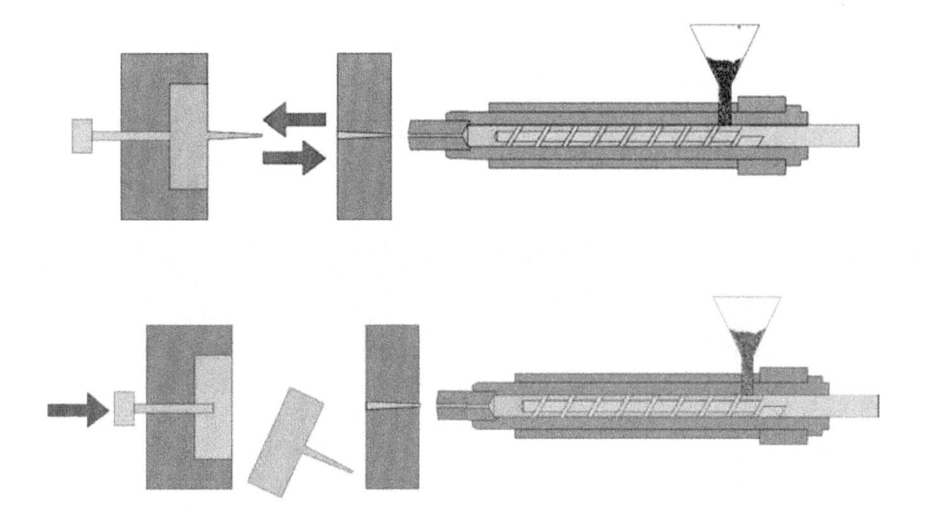

Figure 2.20

Part ejection

Diagram courtesy of ARBURG GmbH

2.1.6.4 Shear damage during processing

The route a plastic material takes through an injection moulding machine is similar to that experienced by a material during extrusion: feeding, melting, homogenisation and forming by a die (extrusion) or mould tool (injection moulding). The major differences are the injection pressure exerted and the higher shear exerted on the polymer. These can have major implications for the material.

As increased temperatures affect viscosity by making polymers less viscous, so higher shear forces also reduce their viscosity, because polymers are shear thinning. This makes the materials more free-flowing, however shear damage can arise during injection moulding, causing physical chain breakage of the molecules and degradation. This is because chain breakage will affect the material properties (see Part 1) such as average molecular weight and viscosity, and physical properties such as strength and impact. Shear can also cause colour changes to materials and reduce the fibre length in glass-filled materials. (Excessive heat can also cause colour changes to occur.) The level of sensitivity to shear varies from material to material. Whilst some materials such as PP are relatively resistant to these effects, others such as PVC will rapidly degrade in harsh processing environments.

2.1.6.5 Materials and settings

Table 2.2 gave a guide to the range of temperatures and pressures used in injection moulding of common materials. It may be interesting to consider these in terms of the melting points and softening temperatures indicated in **Table 1.6** and elsewhere.

2.1.6.6 Injection moulding of thermosets

Thermosets harden in the hot mould tool by crosslinking as opposed to cooling in the mould tool like the thermoplastics. Therefore, they require differing heat regimes. Thermosets for injection moulding can be used in a number of forms: granules, powder, rods or dough. However whatever its initial form the plastic must be able to flow, be plasticised quickly, resist premature curing in the cylinder, and then cure relatively fast once in the mould tool to produce a high quality component.

A typical thermoset injection moulding machine differs from the thermoplastic one described previously in the following areas:

- A specialised thermoset injection unit is used instead of a thermoplastic one. This has a different screw configuration to minimise shear and mixing, however it is still contained in an injection cylinder which houses the heating elements.

- Dosage delay control is also used to minimise material times in the barrel. This delays feeding new material into the injection moulding screw after the mould tool has been filled until the last minute, therefore delaying exposure to heat.

- A mould blow unit is required to closely control the mould tool temperature and gives added cooling control. Venting control and specialised cleaning equipment are also required for thermoset injection moulding operations.

- In contrast to the temperatures shown for thermoplastics in **Table 2.2**, the barrel temperatures on the injection cylinder are now much lower, in the region of 50-115°C. The mould tool is at a higher temperature, typically 150-240°C. Insulated platens are also needed to protect the cylinder from the heat of the tool.

Figure 2.21 illustrates a typical thermoset temperature profile for thermoset moulding.

2.1.6.7 Multi-layer moulding

Multi-layer moulding allows the properties of multiple materials to be incorporated during one process cycle. This presents considerable cost savings as it eliminates further processing steps (e.g. assembly). The multi-layer could take the form of an insert which is placed in the mould tool prior to moulding – this technique is often used to produce decorative mouldings or to over-mould onto metal components. Multiple materials can be injected into the same cavity through the use of several extruders. This can be done simultaneously in co-injection (dual injection) moulding or sequentially in over-moulding. Over-moulding is commonly used to produce the toothbrushes that contain more than one colour and/or material, as well as items such as the calculator keypad shown in **Figure 2.22**.

2.1.6.8 Foamed moulding

Foamed mouldings can be produced by adding foaming additives (blowing agents) to the plastic prior to injection. Gas releasing additives can be used or gas can be directly applied into the melt stream.

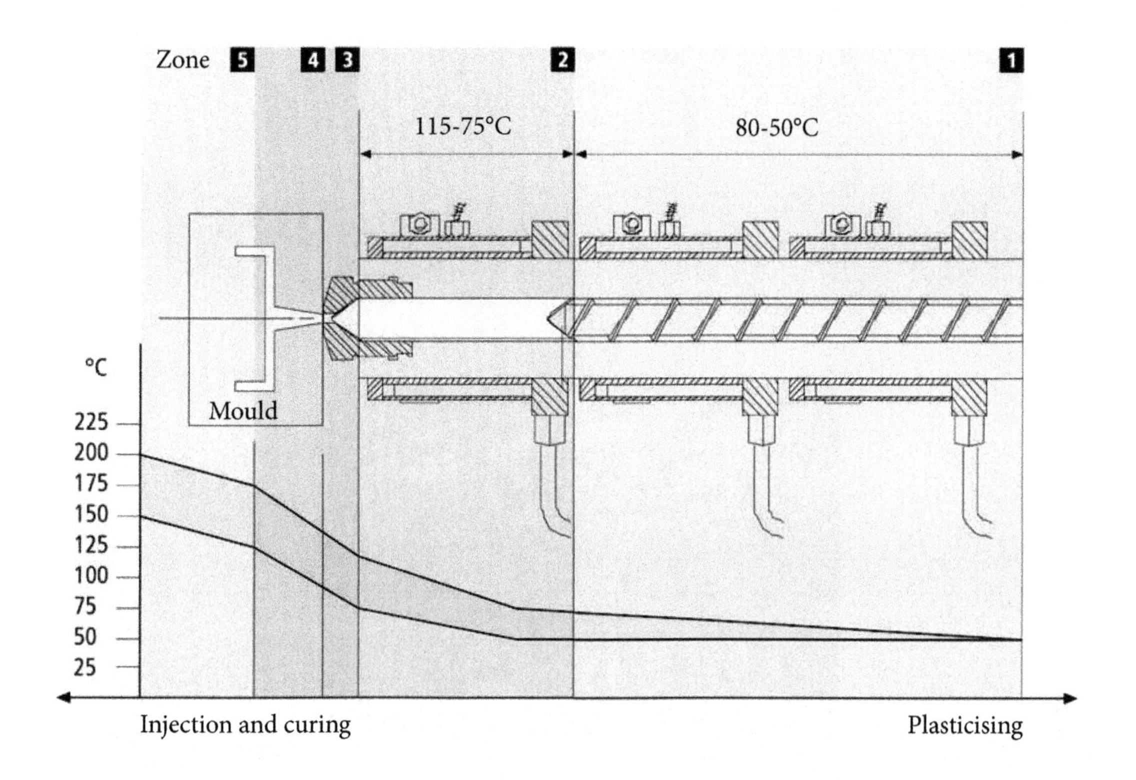

Zone		
Zone 1	Pre-heating zone	Effect of screw geometry, rotational speed, back-pressure and cylinder temperature control
Zone 2	Plasticising zone	
Zone 3	Friction dependent upon injection capacity (injection pressure and injection speed) and nozzle bore	
Zone 4	Friction caused by sprue and gate cross section	
Zone 5	Effect of mould heating and curing	

Figure 2.21

Temperature profiles for thermoset injection moulding

Diagram courtesy of ARBURG GmbH

Figure 2.22

Two colour moulding of keypads
Drawings courtesy of ARBURG GmbH

A range of sizes of cellular structures can be produced depending on the type of process machinery and additives used, including microcellular sized foams.

2.1.6.9 Gas assisted injection moulding

Gas assisted injection moulding (GAIM) is used to produce hollow parts. This works by injecting only a partial shot of plastic material (not enough to fill the mould tool cavity). This is then forced and packed into the cavity by an injection of high pressure gas which is generally nitrogen. The advantages of this technique are weight saving and also a reduction in cycle time due to the cooling effect of the gas. An application where this technique may be applied is in an automotive door handle. A newer version of this technique uses water instead of gas – water assisted injection moulding (WAIM).

Using gas and water assisted injection moulding hollow articles are produced using the injection moulding process route. For the next process, being hollow is a basic requirement associated with the use of the technique.

2.1.7 Blow moulding

The third most commercially important process for plastics production after extrusion and injection moulding is blow moulding, and this process originally developed based on glass blowing techniques in the 1930s. However it was not until the 1940s and the introduction of LDPE that blow moulding offered any commercial opportunities in a market dominated by glass. This breakthrough came in the unlikely form of 'squeezability' – a property which a rigid material like glass could not hope to match. In the 1950s, the introduction of HDPE and commercial blow moulding equipment formed the basis of the blow moulding industry today.

While polyethylenes dominate the blow moulding industry, along with polypropylene, engineering plastics such as nylon, polyurethane, thermoplastic elastomers and blends such as PPO/PS and nylon/PET play an increasingly important market role. This process is only suitable for thermoplastic materials and it is used to create hollow articles such as fuel tanks, milk bottles, beverage containers, etc.

There are two main variations on the process:

- Injection blow moulding
- Extrusion blow moulding

In both processes a tube-like structure called a **preform** is created which is then inflated against the sides of a blow moulding tool and cooled.

In injection blow moulding the material is *injected* to form a tube-like preform, and in extrusion blow moulding the material is *extruded* to form a tube-like preform. The basic process is then the same for both: the preform is inflated to the shape of a mould tool cavity, then cooled, before the tool is opened to eject the part.

Blow moulding materials must exhibit very specific properties. Firstly they must be of a suitable viscosity. They must also have a high melt strength. Examples of both extrusion and injection blow moulded items are shown in **Figure 2.23**.

2.1.7.1 Extrusion blow moulding

In extrusion blow moulding, unlike conventional extrusion, the extruder is vertical and the extrudate (called a **parison**) hangs downwards under gravity from the die. This requires specific properties from a blow moulding material. If the melt strength is weak the parison will not be formed, and the plastic will simply fall on the floor.

Once the parison reaches a preset length the blow moulding tool will close around it. This is a two piece mould tool as shown in **Figure 2.24**.

Another factor in blow moulding is that to produce a container from a tube it is necessary to seal the base of the container. Therefore a further material requirement is the ability to form a seal during the

Figure 2.23

Containers produced by blow moulding
Source N. Goodship

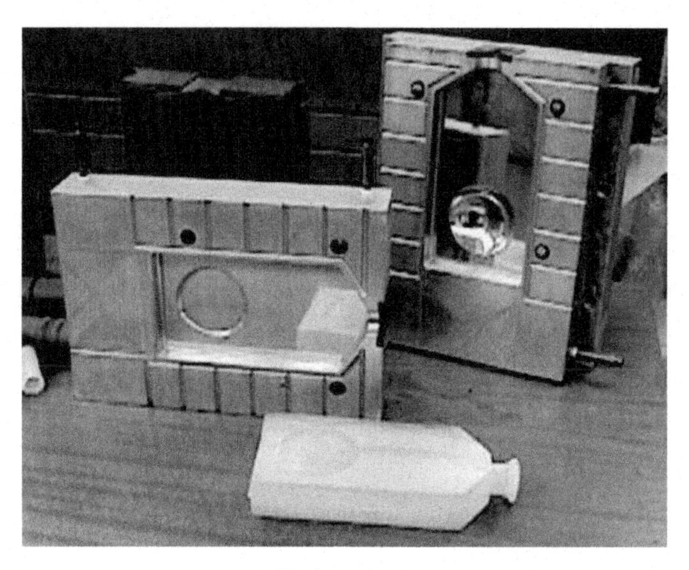

Figure 2.24

Extrusion blow moulding tool and part
Source N. Goodship

forming process. Moulds are fitted with a **pinch-off** area which forms a seal. Containers created in this manner can therefore be recognised by the weld line on the base as shown in **Figure 2.25**.

Once enclosed in the mould tool and sealed a **blow pin** is inserted into the tool which adds compressed air to inflate the plastic against the cavity walls. If the material has enough melt strength to form a parison, but not enough elasticity, it will break or split when subsequently inflated.

Figure 2.25

Extrusion blow moulding showing the characteristic weld line on the base

As expansion occurs from air injected from a blow pin into the mould tool cavity, this means the mould tool must first be moved away to a blowing station to allow continuous extrusion of the next parison, and this is done automatically by the machine. Blowing air extends the material to match the geometry of the cavity as shown in the sequence in **Figure 2.26**. The parison size must be longer than the part and therefore excess material is produced during each cycle here. The pinch off unit at the bottom of the tool both seals the base, and allows easy removal of excess material at the bottom (called flash) once the part is ejected. Therefore extrusion blow moulding by its nature produces scrap material from each component. However this can be collected, reground and reused in the same process.

Blow moulding pressures (5 bar), are much lower than those used in injection moulding (several hundred bar) therefore aluminum moulds can be used. (In injection moulding steel is commonly used which is more expensive and harder to machine.) The most commonly recognised extrusion blow moulded items are plastic milk bottles which are made from HDPE (**Figure 2.27**). However extrusion blow moulding can also be used to make very large barrels and tanks, or much smaller hollow parts around the size of ballpoint pens.

2.1.7.2 Multi-layer blow moulding

In multi-layer blow moulding instead of using a single extruder to produce a parison, a number of separate extruders are fed into a die head, which combines the materials into a multiple layered parison (see coextrusion for a similar process). This could be just two layers, or as many as six or even more. In this way it is possible to produce extrusion blow mouldings consisting of a number of

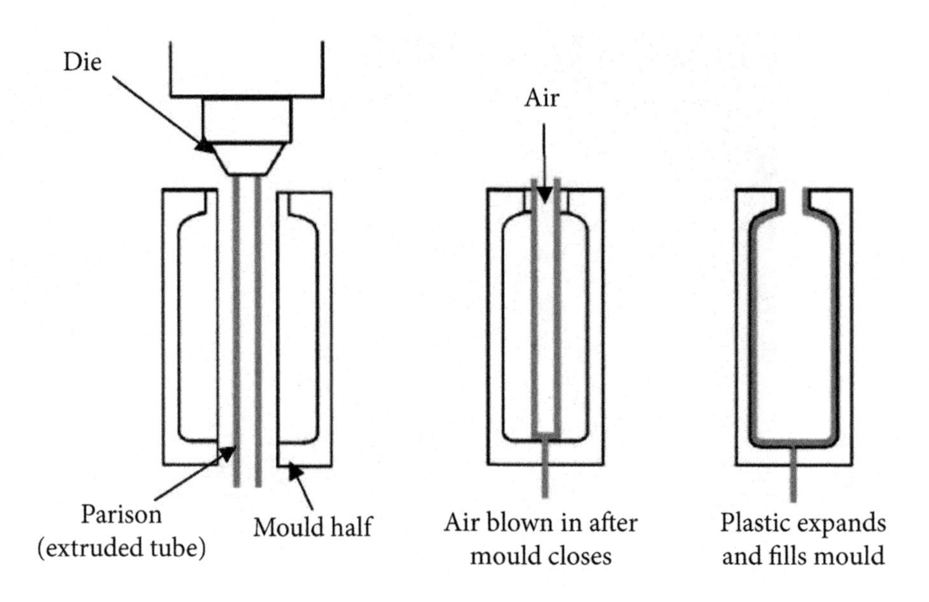

Figure 2.26

Tube inflation in the extrusion blow moulding process

Figure 2.27

HDPE milk bottles
Source N. Goodship

material layers. For example ketchup bottles can be made of five layers comprising: HDPE/adhesive/EVAL/adhesive/HDPE. The inner and outer surfaces (those that can be seen), consist of just HDPE. The EVAL is a material that provides an oxygen barrier to preserve the contents, and the adhesives ensure that all the layers stick together, as HDPE will not stick to EVAL on its own.

Using a similar process, but different materials, 90% of new cars manufactured in Europe have blow moulded plastic fuel tanks. The HDPE material used for this is ultra-high molecular weight HDPE. Six-layer structures improve the fuel impermeability by incorporating barrier layers and adhesive layers (also called tie layers), as well as layers to incorporate process scrap:

HDPE/adhesive/nylon/adhesive/recyclate/HDPE

The nylon material acts as the barrier to petrol vapour, and again, the adhesives ensure that all the layers stick. Nylon will not stick to HDPE on its own.

2.1.7.3 Injection blow moulding

Injection blow moulding is most commonly used with poly(ethylene terephthalate) - PET to produce beverage containers. However LDPE, PP, HDPE, PS, PVC and PAN - poly(acrylonitrile), can all be processed this way.

As the preform is created by injection moulding and gated at the base, there is no weld line visible on finished parts as there is on extrusion blow moulded products. However, there may be a clear central mark visible at the centre of the base which indicates where the plastic was injected into the mould tool cavity, this remnant is a gate mark as shown in **Figure 2.28**.

Figure 2.28

Injection blow moulding showing gate mark

Because there is no issue with a sealing weld line in injection blow moulding, high pressures can be used for inflation that could not be tolerated by extrusion blow moulding without bursting of the parison. The downside is that because of the two part manufacturing process used, with separate injection and inflation steps, there are higher tooling costs. There are also design limitations, for example features like handles cannot be incorporated, unlike extrusion blow moulding.

A great advantage over extrusion blow moulding is in the level of scrap generated. In this case there is none. The parts can also be produced to a very high and consistent quality, and as they are injection moulded first there is no variation in bottle weights.

Injection blow moulding is carried out in three distinct stages: first a preform is produced by injection moulding. This looks similar to a test tube. Inflation is the second stage to produce the bottle or inflated component (**Figure 2.29**). Thirdly the part is removed. In extrusion blow moulding the process cycle is cyclical and carried out on one machine, very much like injection moulding. One stage cannot commence until the previous stage finishes. In contrast, in the case of injection blow moulding, the three distinct phases can take place all at the same time, with three different components on a different stage at any given time. This can all be done on just one machine, or preforms can be produced for inflation on another machine at a later date. This is because the injection and inflation stages are distinct processes and can be separated.

When making PET bottles, the preform is injection moulded with a residual wall thickness of around 4 mm. If a low temperature is used on the injection mould tool then crystallinity (see Part 1) can be kept very low. When the preform is subsequently heated (generally with infra-red heaters) to 100°C ready for inflation (where it is in a rubber-like state), it is shaped in two ways.

1. It can be stretched in length (axially) by a rod inserted through the neck called a core rod (as seen in **Figure 2.29**). This is also used to transfer mouldings from the moulding station to the inflation station.

2. It is inflated off the core rod by an air supply to the dimensions of the mould tool cavity. This creates an inflation shape, similar to blowing up a cylindrical balloon. There is no inflation of the neck and the base expands less than the sides.

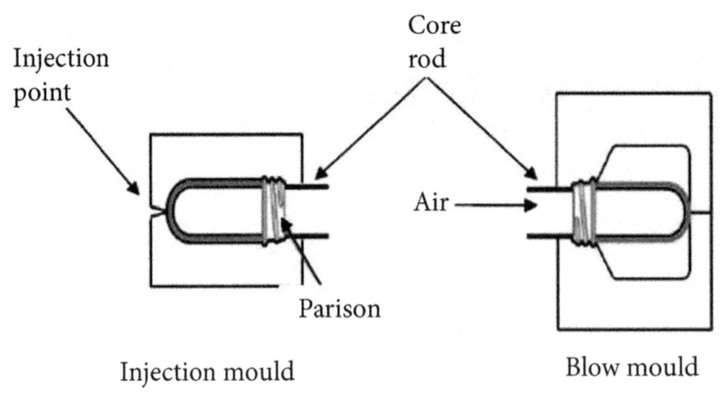

Figure 2.29

Injection blow moulding process

As crystallisation occurs in PET when an extension ratio of 2 is exceeded, in injection blow moulded PET bottles there will be both crystalline and non-crystalline regions depending on the levels of expansion needed to fill the tool. An interesting way to study crystallisation in PET bottles is to heat them to 120°C for 10 minutes. The neck and base will crystallise and go opaque while the wall will both shrink and remain transparent. This is because the crystals in the bottle wall are highly orientated in the direction of expansion whilst other regions are barely inflated and amorphous.

2.1.8 Film blowing

Film blowing is used to create thin plastic films. This product is then used to make many familiar products like carrier bags, dustbin liners, plastic sacks and plastic wrappings. The plastic feedstock in the form of pellet or powder is fed into a material hopper where it is gravity fed onto an extruder screw. The rotation of the screw draws the material through the heated extruder barrel which creates a molten plastic as it plasticises and mixes it. At the end of the extruder is an annular die opening which creates a tube of molten plastic material. As air is then blown up the inside of the molten tube, the plastic is inflated, whilst an air ring on the outside cools the plastic. The plastic expands evenly to create a bubble. The bubble is pulled upwards onto rollers where it is 'nipped' to close the bubble and then directly wound onto collection rolls. This process is illustrated in **Figure 2.30**. There is no waste as the process is continuous once it is set up. A consistent thickness (called the gauge) and width can be achieved.

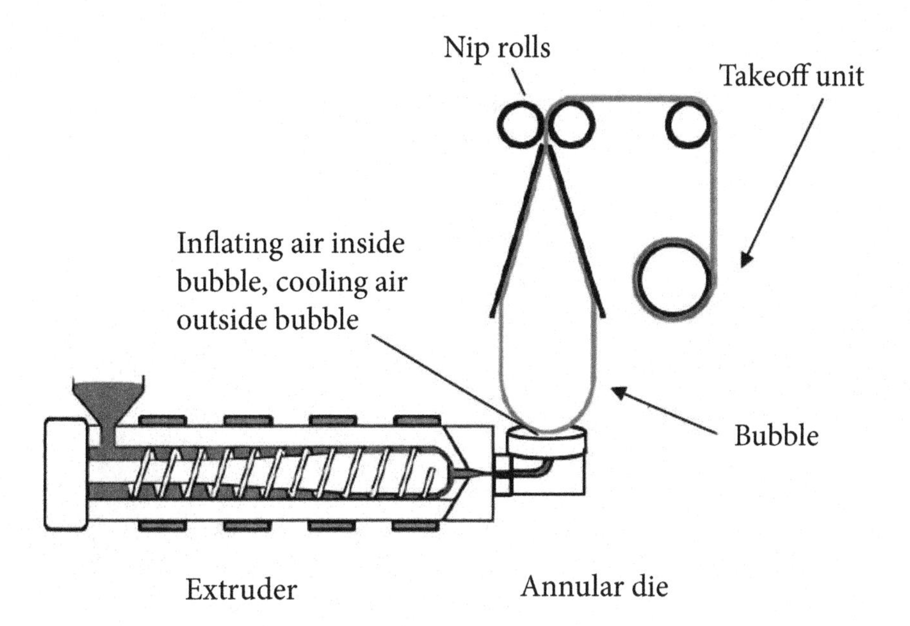

Figure 2.30

Blown film process

The plastic feedstocks used must have the ability to inflate evenly and have sufficient melt strength. Therefore materials must have a suitable viscosity for this process. Common materials used are those associated with the packaging industry and the majority are polyolefin materials such as low density polyethylene (LDPE), linear low density polyethylene (LLDPE) and high density polyethylene (HDPE). The blow up ratio (that is the diameter after inflation compared to diameter before inflation) is usually in the range of 2:1-4:1.

2.1.9 Thermoforming

In thermoforming the feedstock material is a sheet as opposed to pellets or powder, and thermoforming is therefore termed a secondary shaping process. The term thermoforming is applied generically to the technique of shaping a sheet, and encompasses a number of variations performed using vacuum, pressure or drape forming. Thermoforming has been around since the Egyptian Pharaohs who formed pre-heated tortoise shells into cooking utensils. Luckily for the tortoises we have plastic and metal ones now. More recently in World War II, where a number of plastic materials found new uses, thermoforming was used to produce the plastic canopies of aircraft. Today thermoforming is used to produce blister packs, ready meal food packaging, meat trays, yogurt containers, food insert trays (for example the ones that position sweets in a box of chocolates) and margarine tubs and lids amongst many others. Larger thermoformed components can include baths, shower trays and housings for example. **Figure 2.31** shows some components made by both thermoforming and injection moulding, the injection moulded ones can be distinguished either by the visible sprue mark or the complex geometry.

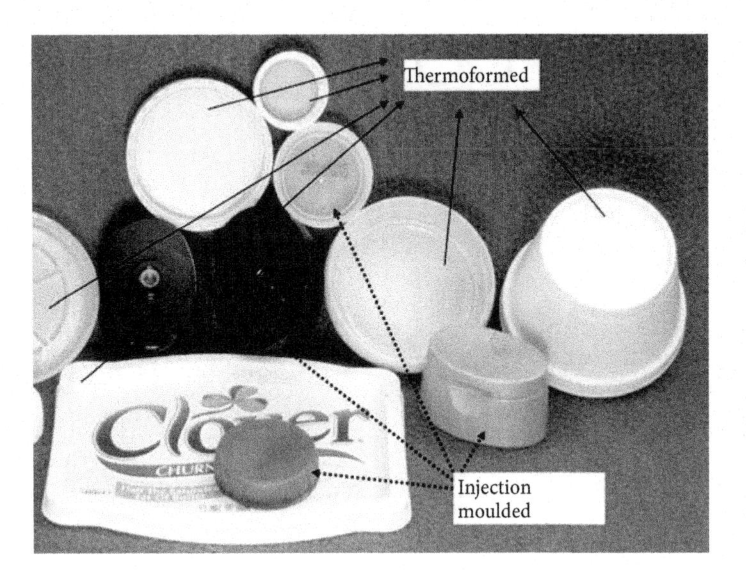

Figure 2.31

Some common thermoformed and injection moulded packaging items (left) and Darth Vader thermoformed mask (right)

A variety of thermoplastic (not thermoset) sheet materials can be used such as PVC, PP, PS, ABS, PC, PMMA, PET and Teflon PTFE for example.

As the range of products described above suggests, thermoforming is used to create curved parts and cavities such as pots, tubs, containers, bases and trays from flat sheets. The sheet feedstock material can be of varying thickness, however because of the need to obtain a consistent heat throughout the thickness of the sheet, most products will be made using sheets under 5 mm thick. The sheet is clamped into a thermoforming machine and is then heated on one or both sides usually by infra red heaters. For thicker sheets (for example, typically 6 mm PMMA is used for thermoforming baths) ovens are used for pre-heating. For thinner sheets of 2 mm or less, direct heating with radiant heaters over the sheet can be used. The heating level used will depend upon the thickness of the sheet.

In sheets thinner than 0.5 mm the temperature gradient from top to bottom will be small. However for thicker sheets, the entire thickness of the sheet must be at the same temperature, and this must be achieved without overheating the surface. The heat source may be cycled on and off to allow the temperature to equalise, or heating may be provided by two sets of heaters above and below the clamped sheet.

In the simplest version of a thermoforming process, the sheet is heated and then lowered onto a concave (female) mould tool under vacuum as shown in **Figure 2.32**. As the sheet contacts the mould tool it is stretched and formed by vacuum into the shape of the mould tool. This stretching and forming action performed during any thermoforming process is referred to as drawing. The draw ratio can be simply defined as mould tool depth/width. The deformation time is low, as once contact is made with the tool surface the plastic will solidify. Hence it is the heating of the sheet that dominates the cycle time in this process.

One disadvantage of forming into a female concave cavity is the excessive thinning that can occur at the corners. This occurs because freeze off on contact is so rapid that the areas that require more stretching (typically the corners) actually have less molten film to play with. The thinnest part will always therefore be the bottom corners. This can often be seen in coloured packaging products such as box inserts if held up to the light, where the corners will be far more transparent as they are thinner. Indeed this is an indicator that a product has been formed by this method.

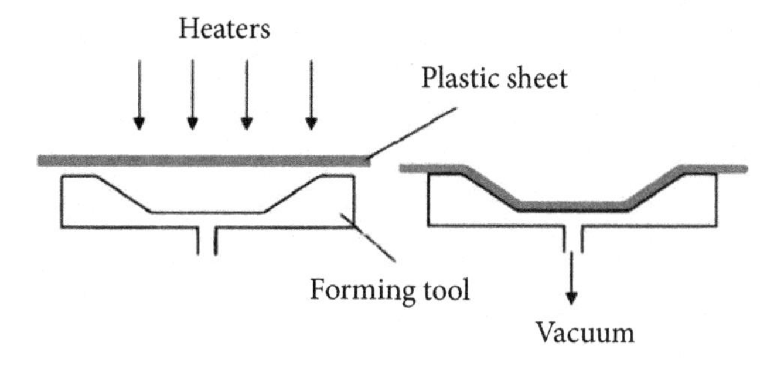

Figure 2.32

Example of thermoforming (vacuum forming onto a female mould tool)

By varying the draw ratios of the cavity, various plastics can be evaluated for their ability to form. Increasing the material temperature tends to increase the potential draw ratio, as this decreases the viscosity of the material, but any given polymer will still have a maximum achievable draw ratio.

For PVC for instance, a material which is frequently thermoformed, forming temperatures of between 60°C and 100°C are used (the Tg of PVC is 68°C). Above 180°C thermal degradation occurs rapidly.

To overcome the thinning at corners, a convex (male) mould tool can be utilised. By inverting the process, the mould tool can be used to stretch the side walls of the container. Examples of items manufactured this way are vending machine cups and margarine tubs. In this method the base thickness is thicker than the wall thickness and thin containers can be produced. However if the depth of the container is too great in proportion to the width, blow moulding would be more suitable.

Many variations of thermoforming exist, as forming can be carried out by both pneumatic and mechanical means and can include pre-stretching stages. Vacuum or compressed air can be applied, and matching male and female moulds can be closed onto heated sheets. Mechanical plugs can also be utilised to assist the forming process. Automation of the process is common. Machines can be fed continuous rolls of sheet which are passed after forming to automatic cutting devices (or cut in situ) to trim waste sheet and separate individual items. Sheet waste can be reclaimed and incorporated into the new sheet material in many cases.

Multi-layer sheets can also be formed in this way, as can composite materials such as glass-filled polypropylene sheets which are used in the automotive and aerospace sectors.

A whole range of plastic materials can be formed as long as they retain sufficient strength not to burst or split when elongated during thermoforming. Some materials can be stretched a great deal (600%), others may manage only 10%. Obviously this limits the shapes that can be achieved using different material types.

2.1.9.1 Mould tooling for thermoforming processes

As thermoforming is a low pressure process, the demands on the mould tooling are much lower than for forming dies in extrusion, or the mould tools in injection moulding or blow moulding. Mould tools produced using aluminium are therefore adequate, and they also have higher thermal conductivity than steel, which allows better temperature control and cooling. Mould tools can even be prototyped using wood, epoxy or polyester materials, and these may also retain enough strength for short production runs.

There are no high pressures and forces involved in production and therefore the capital costs associated with thermoforming are low compared with injection moulding for instance. However the higher cost of using an initial feedstock in the form of a sheet must be taken into account. The moulding will also generally need to be trimmed to remove excess sheet material and therefore there can be a large amount of process scrap.

In designing for thermoforming, the wall thickness decrease in relation to the increased area of the new component must be kept in mind. Evacuation of air though vacuum or air vents in the tooling is also necessary for efficient drawing as fast vacuuming or drawing is an essential feature.

Disadvantages of thermoforming include the inability to include features such as reinforcing ribs, so the product is likely to buckle under compressive forces. One way to overcome this is to use corrugations. However thermoforming is limited to fairly simple shapes and products such as trays and cups.

2.1.10 Compression moulding

Compression moulding can be carried out on both thermoplastic and thermoset materials, although it is far more commonly used for thermosets. The most commonly used material in this process was developed in 1908 by Dr Leo Baekeland as Bakelite, which is phenol-formaldehyde resin.

The process itself is relatively simple (**Figure 2.33**). A two piece mould tool is heated. When opened, a specific weighed amount of the desired material is placed in it. It is then closed under pressure, compressing the softened material. During this compression the material reaches sufficient temperature and pressure to flow and forms to the shape of the mould tool cavity. Once it has solidified sufficiently to retain its strength, it can be ejected.

The cavity depth of a compression moulding is an important parameter for the moulding pressure required. Tooling can be heated by electric cartridge heaters (most common) or oil or water heaters. The tool consists of two halves, one moving, one stationary, which register together on closing by the force provided by mechanical, pneumatic, hydraulic or electric means. The mould tool is fixed onto platens, as with injection moulding. **Tie bars** provide for very precise alignment when the mould halves are bought together under pressure. Compression machines are rated by their closing force (clamping capacity) and a wide range of machines are available from 6 tonnes to 5,000 tonnes or more.

Material can be either fed cold or preheated into the tool. Preheating reduces curing times in the case of thermosets and is always likely to be necessary for thermoplastic materials. When the tool is closed and the material compressed it fills out to match the geometry of the cavity. Thermoset materials

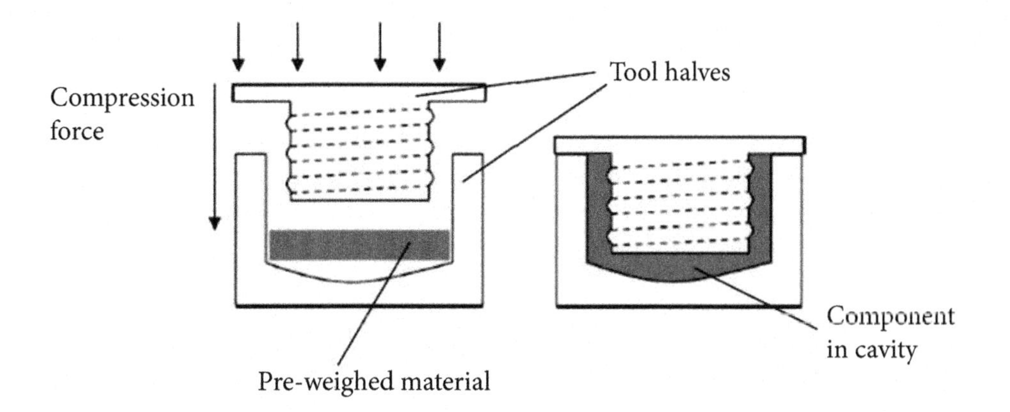

Figure 2.33

Compression moulding

can be removed once cured. As with injection moulding, thermoplastic materials may require some cooling to solidify before removal.

For simple geometries such as computer housings, compression moulding can be an effective production method. Large parts can be achieved, which are only dependent on the size of the press. The machine can also be automated. Significantly for thermosets, the waste from injection moulded articles is avoided (such as sprues), also there is less stress moulded into the component and reinforcement materials are not broken up as they can be during the injection moulding cycle. However, it is not possible to compression mould very complicated parts such as can be achieved with injection moulding, and it is also difficult to achieve the same level of tolerance if high precision articles are required.

There are several critical parameters in this process:

- Design – part geometry (including projected area and depth), wall thickness and flow path
- Speed of closing of mould tool
- Material – specifically the material's ability to flow once compressed
- Mould tool temperature (and the temperature of the compound prior to placing in the mould tool – pre-heating may be necessary)
- Mould tool cavity surface finish
- Moulding parameters such as clamping pressures, mould tool temperatures and cooling times which control the moulding cycle times

In processes with in-built dosing methods such as injection moulding, the shot weight and temperature are controlled. If using automatic compression moulding (continuous automated production without operator input), dosing units can also be employed, however semi-automatic processes (where operator action is required after each cycle) use pre-heated preforms. Using a high frequency dielectric oven, the required temperature for the preform can be reached in seconds.

The benefit of pre-heating is to overcome the low thermal diffusivity as this process requires heat transfer. For thermosets, crosslinking can be time consuming, and cycle times in excess of several hours are possible. In the case of thermoplastics, the tool needs to be cooled before the part can be ejected, and again the cycle times can be as long as 12-20 minutes.

This method is generally only used for relatively thick sections and reinforced materials such as GMT. Compression moulding usually produces relatively simple shapes with few features and simple geometry. However in contrast to injection moulding, the shear levels experienced within the tool are relatively low.

2.1.11 Injection compression moulding

As the name suggests, injection compression moulding incorporates aspects of both the compression moulding and injection moulding processes. This is illustrated in **Figure 2.34**.

A pre-set amount of material is metered in an injection moulding cylinder and injected into a partly closed mould tool. The mould tool is then closed and compression is used to complete the forming of the part in the cavity. Filling pressures are substantially reduced with this process as compared to injection moulding and there is less orientation of the material, as it is subject to squeeze flow (compression) rather than fountain flow (injection). Material does not fill the mould tool cavity in the usual manner as with a conventional injection moulding system. For thin walled and transparent mouldings this can be an advantage.

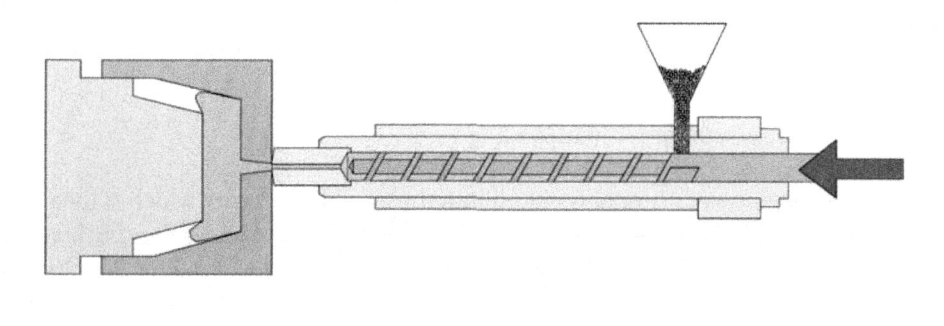

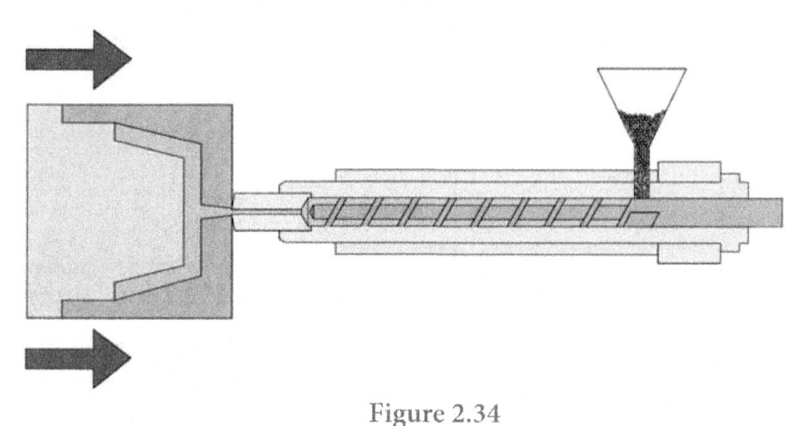

Figure 2.34

Injection compression moulding
Diagram courtesy of ARBURG Gmbh

2.1.12 Rotational moulding

For the production of hollow shapes, rotational moulding offers an alternative to blow moulding. This technique is also commonly known as rotomoulding. (A very similar process to this is used in food processing for making hollow chocolate shapes.)

The process uses a powder or liquid raw material placed inside the mould tool cavity, which is then rotated biaxially and heated. Gravitational force ensures that the powder becomes evenly distributed on the hot mould tool cavity walls. It should be noted that centrifugal force plays no part in rotational moulding.

In order to obtain this distribution, two axes of rotation are required. These are in the central planes horizontally and vertically. The minor axis of rotation controls left to right (or right to left) rotation. The major axis is in the up/down direction.

The mould is moved between a number of 'stations', and the process can typically be characterised by the four steps illustrated in **Figure 2.35**:

1. Loading of material into the mould tool cavity
2. Moulding under rotation
3. Cooling of the part in air
4. Cooling of the part in water prior to unloading

First a pre-weighed amount of material is placed into the mould tool cavity which is then clamped shut. The mould tool is moved into an oven and rotated. The rotation speeds may vary but are generally different for the two axes of rotation. A rotation speed ratio of 4:1 (minor: major rotation speed) can be used for symmetrical objects, for example rotation speeds of 40 rpm minor and 10 rpm major

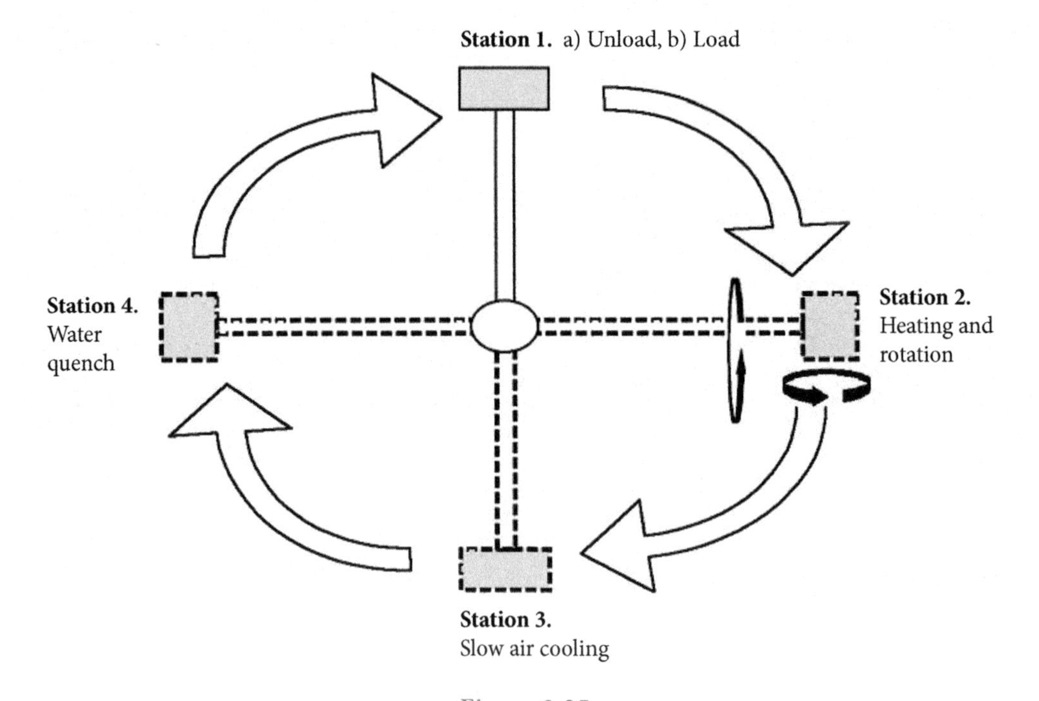

Figure 2.35

Rotational moulding process

– although this is a high-speed example. However for parts with unsymmetrical configurations, a tailored ratio of rotation would be required. For example, a component may require more material in one particular region, to enable a thicker section to be achieved.

As the oven heats the mould tool, the powder or liquid will begin to melt and will eventually all form a molten layer in the cavity, distributed evenly by the rotation of the mould tool. Once all the raw material has been distributed in this way and the final component has been formed the moulding needs to be cooled. This is done in a cooling chamber and can take the form of a cold water spray, cold air jet or a cold liquid circulating through the mould tool (similar to those methods used in other processes such as injection moulding and blow moulding).

The final stage of the process is part removal from the cavity. This can be done manually or again by methods employed in other plastics shaping processes such as mechanical ejection or air ejection.

Compared with processes such as injection moulding and blow moulding, the cycle times used for rotomoulding are very long, typically up to 15 minutes. However cycle times may be under 5 minutes or even as long as 30 minutes, depending on the size and wall thickness required in the final component.

Rotomoulded components are relatively stress-free as a result of the low pressures used, and in addition the process tends to encourage material to build up in corners, where more strength is required in the components. This contrasts with blow moulding for example, where the corners are typically thinner than the rest of the moulding. In comparison with blow moulded products the strength is therefore generally superior due to a more uniform wall thickness. Rotomoulding also allows for more design variations such as complicated curves, as well as improved control of surface finish – matte, high gloss and all finishes in between can be produced. In addition the scrap generated in blow moulding does not arise in the rotational process.

Rotomoulding has been used for the production of boat hulls, rubbish bins, furniture, toys and light globes. Parts as small as golf balls or as big as giant agricultural tanks (85 m^3) can be produced with this technique.

Other advantages of rotomoulding include the following:

- Low prototyping and tooling costs
- Simultaneous processing of both multiple parts and multiple colours is possible
- Quick mould tool changes
- Metal inserts can be moulded in if required
- Moulded-in inserts and graphics can be used
- Double walled parts can be moulded as well as multiple layers
- No scrap is produced

Whilst most thermoplastics can be processed using this method, the most widely used material is polyethylene. Other materials utilised by the industry include poly(vinyl chloride), nylons, polycarbonate, polyesters such as PET and PBT, ABS, acetal and cellulose type materials. It should be noted this list is not exhaustive. Many other materials including thermosets such as phenol-formaldehyde, epoxy and silicone materials can also be used in this process.

Moulds are generally low cost compared to blow moulding or injection moulding, with cast aluminum being the most widely used material for construction. Multiple cavity moulds are generally used – except where the large size of the component limits this use. Dimensional tolerances are around 5%, which is a function of the materials' shrinkage. Wall thickness can be adjusted by the amount of powder charged and cycle times. Whole areas of a mould tool can be kept cool to prevent plastic formation in those areas as required. For example, a lidless hollow container could be formed by leaving an insulating disc – and therefore a cold mould tool surface – at one end.

As mould tool cavities are female, surface detail can only be controlled on the outside of the moulding – the inside of the part contacts only air. So for components requiring inside detail this process cannot be used and injection moulding would be more applicable. This also applies if very high surface detail is required, where the high pressures employed in injection moulding produce better reproduction.

2.1.13 Calendering

Calendering processes are used in the production of plastic film and sheet (as an alternative to extrusion, which can also be used). Calendering is a continuous method which fuses a layer or a series of layers into film or sheets of either rigid or flexible materials. The feedstocks used in calendering are similar to those used for compounding. The raw materials are blended to the required formulation and then fluxed and heated to fuse them together.

In this process the plastic, which is now in the form of a molten paste, is compressed between two heated cylinders which rotate in opposite directions. The small gap between them is called the nip region. The plastic is squeezed into the form of a sheet, which is subsequently drawn down into thinner and thinner gauges as the process proceeds. There can be a number of calender configurations, using two or more banks of rollers. Once in the nip region, the rotation causes an increase in frictional heat and therefore material temperature due to the high velocities used. However, the time scales involved are short and degradation of heat sensitive materials can be avoided.

Widths of up to six feet are achievable. The width of the material is significantly increased as it exits the nip. Film thicknesses as fine as 0.15 mm can be produced at speeds in excess of those possible with extrusion. PVC is the most widely prepared material using this process, including flexible and rigid sheets and films. ABS is also commonly processed this way.

This process can also be used to change the surface appearance (surface finish) of a preformed sheet or to laminate materials together if embossed rolls are used. This is transferred to the sheet as it passes through.

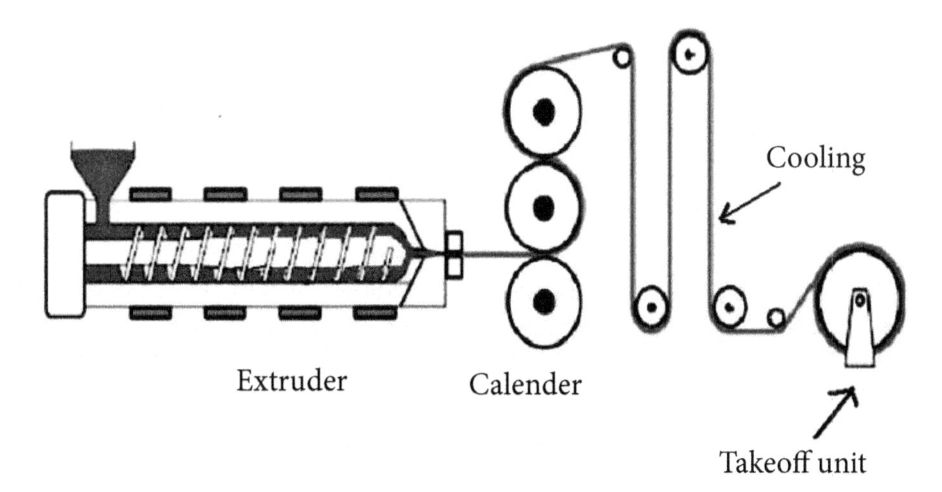

Figure 2.36
Calendering process

Common products made by this process include vinyl floor sheets and coverings, and decorative films. A schematic of the process is given in **Figure 2.36**.

2.1.14 Intrusion moulding

Intrusion moulding is not widely used, but the process has specialist applications, and will be described briefly. However before detailing the process it is necessary to clarify two terms: extrusion and intrusion. Extrusion is a forward turning of a screw to feed material, whereas intrusion is the opposite – a backwards movement of the screw to feed material. (In fact, a period of intrusion occurs in every injection moulding cycle as the next shot is dosed, and the screw moves backwards to position.) However, with intrusion moulding, unlike injection moulding, the screw is still turning as the feeding cycle occurs.

The original invention of injection moulding had many aspects of intrusion moulding. These disappeared with the development of the reciprocating screw seen today. Intrusion is therefore not commonly seen, but it is used for mixed plastics and contaminated feedstocks.

In intrusion moulding, a plasticating unit is used to plasticise and homogenise a finely ground mixture which is then intruded into a mould tool. (Plasticating describes the action of the unit as it converts the feedstock into a plastic material that can be shaped and formed.) Open mould tools are commonly used but this process can also be used to fill a closed mould tool, however it is not as effective as high pressure injection moulding, generating only small exit pressures as seen in extrusion. Once cooled the part can be ejected or removed, depending on the mould tool design. This method is especially suitable for mixed and contaminated feedstock as long as the materials are finely ground. The small size of the particles allows them to pass through the plasticating unit. A variety of materials such as sand, glass, paper and wood can be incorporated as long as there is a minimal amount of plastic to use as a flow base. This is generally found to be around 40% of a low melting point polyolefin fraction.

This method can be used to produce simple, large and thick mixed plastics components that can be used as non load bearing wood replacement products such as pillars and posts. However, the properties of the final components are not particularly good. This technique has only limited use, usually for reusing waste materials.

2.1.15 Transfer moulding

Like injection compression moulding, transfer moulding also utilises elements of both injection and compression processes. However, this process is generally only applicable to thermosets. In transfer moulding material is placed in a cylinder called a transfer pot, and then plunged under pressure into the mould tool when the press is closed. The clamping pressure holds the mould tool shut while the top cylinder exerts pressure to force material from the feeding unit and through the runner system and into the mould tool cavity. See **Figure 2.37**.

The advantage over compression moulding is that only one feed point is needed to fill several different cavities (like injection moulding) as opposed to a feed point for each part with compression moulding. The sizes of the gates and runners that feed the material into the tool depend on the type of material being used. If using reinforcements such as glass fibres, these need to be of sufficient size not to degrade the properties of the final components, and to allow fast cavity filling. Compounds with smaller particle size fillers can use smaller runners and gates.

Compared with compression moulding, when using this method a greater hole depth/length ratio can be achieved. The cycle time is faster and more advanced design features such as undercuts can be included. There is also less flash and less waste with this process route.

Also compared to compression moulding, thinner sections can be produced and tooling tends to be cheaper since there is less pressure on the system. However there are waste materials generated with each shot including that left in the pot, the sprues and runner.

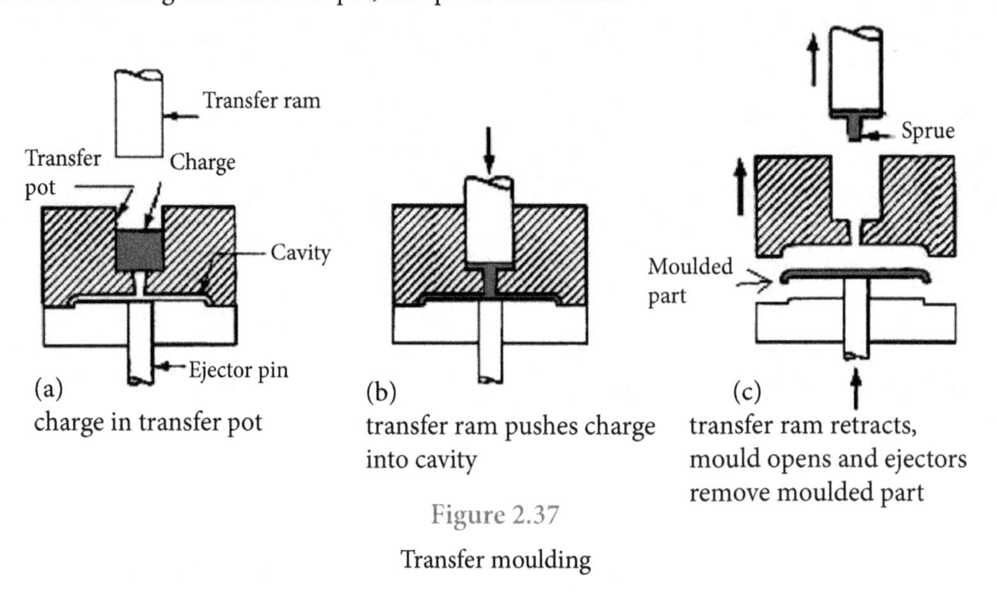

Figure 2.37

Transfer moulding

These machines are used with BMC compounds and also TMC (thick moulding compound, a thickened long fibre version of BMC) to produce components such as electrical switchgear, wiring devices and appliance parts.

Material can be preheated in the transfer pot to reduce cycle times.

2.1.16 Reaction injection moulding (RIM)

RIM and its process variations use liquid resins and are also known as Liquid Transfer Moulding processes (LTM). A two part system of resin and catalyst is mixed and injected into the mould tool. In a typical process, intermediate polymer-making chemicals such as isocyanate and polyol (both prepolymers) are combined to make polyurethanes. Instantaneous mixing of resin streams takes place in a high pressure mixing head at the mould tool entrance point and polymerisation occurs during the mould filling phase. The process is illustrated in **Figure 2.38**. The mixture is then forced into a closed mould tool at relatively low pressure (50 psi). Generally the tooling is vented and relatively low cost due to the low pressures involved, however process control is essential as the system contains low viscosity rapidly curing components. Because of the liquid nature of the material, it is possible to

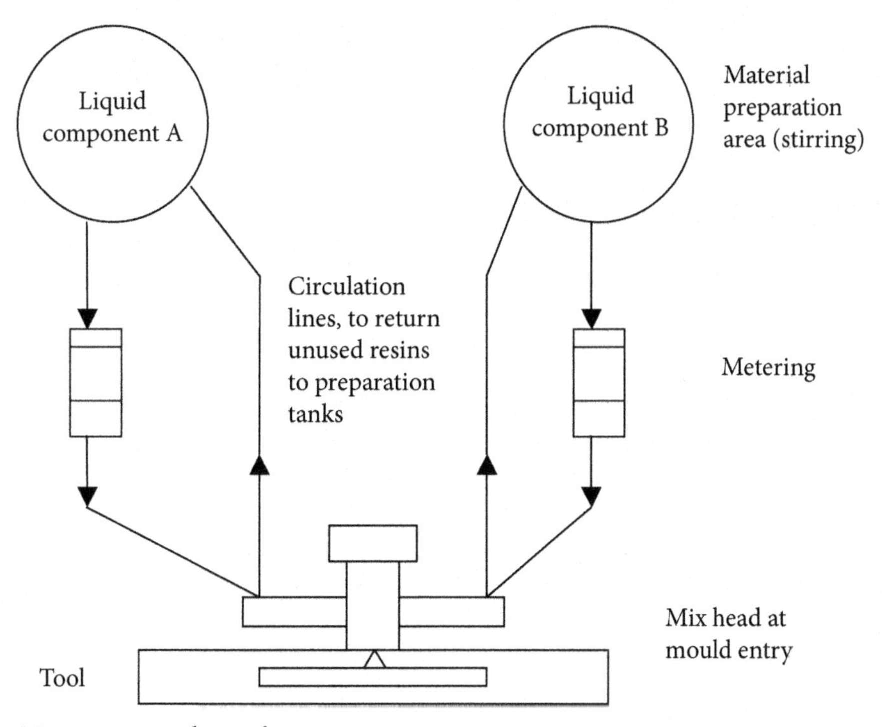

Mixtures are made up of non reactive components.

Liquid component A: e.g. polyols, catalysts + other components

Liquid component B: isocyanate component

Figure 2.38

RIM process

mould relatively complex parts. Although RIM is mainly used with polyurethane, other materials can be processed such as acrylics, epoxies and polyesters. Other variations of this process exist, including the following which are used to make structural components:

- RRIM – reinforced reaction injection moulding, where a chopped reinforcement is placed in one or both of the resin components.
- SRIM – structural reaction injection moulding. Reinforcement i.e. glass mat, is placed in the mould tool before injection and the liquid resin is injected on top. The rest of the process is the same as RIM.
- RTM – resin transfer moulding, also has a mat pre-placed in the tool, however in this case only one catalysed liquid resin is pumped in and positive pressure is used to fill the tool. There is no mix head as in RIM.

2.1.17 Making fibre reinforced structural components

For heavily reinforced materials, where orientation effects caused by processing through a screw are not acceptable, it is necessary to use other processes such as the RIM variants introduced in the previous section. For structural parts, depending on their application, it may be necessary to have isotropic properties (the same in all directions) or to ensure the properties are maximised in one particular direction.

It should be remembered that the load bearing capacity of fibre reinforced materials is enhanced only in the direction of the fibre alignment, and therefore fibre orientation needs to be aligned with the direction of an applied mechanical load. The orientation of various glass products was shown in Part 1.

The following sections will therefore briefly describe the methods used for shaping thermosetting and thermoplastic composites.

Whilst a variety of fibre materials can be used (see Part 1), this section will focus on glass reinforced fibre reinforced plastics (GRP). These are the most widely available materials and have found use in a variety of sectors, including automotive, marine, electrical, aerospace, construction and wind turbines. However, the widespread replacement of metal or timber components in marine applications can give a guide to just some of the wide ranging attributes of these materials.

- Corrosion resistance
- Low density
- One mould tool can produce one entire repeatable component (for example a hull) and replace several production steps with just one
- Reduced maintenance costs

One other property of GRP components is of great relevance to naval personnel and their safety. The lack of metal structures and therefore non-magnetic character of these materials aids in mine clearance operations. The UK launched the first GRP minesweeper in 1972.

Glass fibre reinforced resins have been used for bodywork on automotive applications for a long time. However, they are usually produced in low volumes as special models.

Glass mat reinforced thermoplastic materials are commonly called GMT. They can be made of a variety of materials such as PP and PET and have high levels of orientated glass reinforcement within a polymer matrix. The blanks are heated and shaped by compression moulding. To use materials like this economically, because of the costs involved in producing the sheet materials, it is necessary to ensure production runs are of sufficient length to be cost effective, potentially in the order of 10,000 parts.

For non-structural parts, materials such as BMC and SMC can be used in processes like injection moulding. However for composite production, where structural properties require the inclusion of reinforcing fibres, there are a further two specialised processes of pultrusion and filament winding.

Processes used for the production of reinforced materials can be divided into three categories; manual (hand lay up of fibre glass), semi-automatic (compression moulding) and automatic (pultrusion, filament winding and injection moulding.)

2.1.17.1 Filament winding

This technique is useful to produce highly structural hollow components with simple conical geometries. It is similar to pultrusion (next section) but unlike pultrusion this is a two step process consisting of manufacture and then cure. This technique also differs in that it uses a mandrel for shaping as opposed to a die. A schematic is shown in **Figure 2.39**.

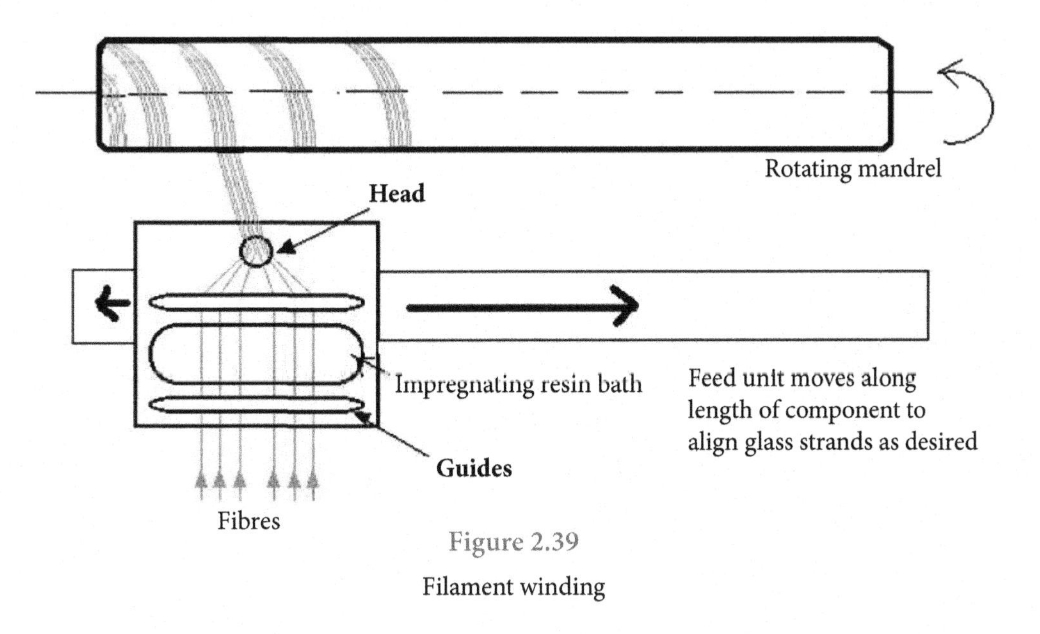

Figure 2.39

Filament winding

Both filament winding and pultrusion use a glass roving (comprising a number of parallel glass strands) which is pulled through a bath of resin under tension and aligned at the specified thickness, geometry and orientation. In the case of filament winding this is then wound directly onto a rotating tool called a mandrel until the desired component shape is achieved.

The most common plastics used are epoxy and polyester. The impregnated glass fibres, which can consist of a number of layers with different patterns and orientations, are fed through a feed head onto the mandrel which forms the shape of the inside surface of the final component. The mandrel is fully sized to the desired component which can range from a small diameter pipe to a larger corrosion resistant pipe or conical tank depending on what is required. Once complete the composite material is heated to cure the polymer. The mandrel can either be removed or remain as part of the final component design.

As long as the mandrel can be made big enough, there are no real size limits restricting this technique, however as the strands are all wound individually, the process gets slower as size increases, unless the number of rovings is also increased. Filament wound products can be used in very specialised and high performance applications such as in the aerospace and automotive industries. Because of the individual control that can be exerted over individual fibres and their orientation, very high levels of reinforcement can be achieved with this technique.

2.1.17.2 Pultrusion

Compared with filament winding, which is a similar technique, pultrusion incorporates the cure stage within the process. It is used to make the structural versions of the types of components that can be produced by thermoplastic extrusion, such as structural platforms, walkways, supports for pipes, trusses, bridges and handrails. Again, the major reinforcement used is glass fibre with unsaturated polyester resin the most commonly used plastic. The glass strand(s) is fed through a resin bath, which impregnates the glass with resin, and excess resin must be scraped off before entry to the forming unit. Here the material is aligned as required, and then passed into a heated die where the reinforced resin is shaped and cured. Like extrusion, the final cross section is the result of the shape of this die. The heat in the die causes the material to cure, meanwhile it is being constantly pulled through by the pulling devices at the desired speed for curing. The process can therefore be split into four stages: impregnation of glass with resin, consolidation, cure and take off. A schematic is shown in **Figure 2.40**.

Profiles are cut to length as required using downstream equipment.

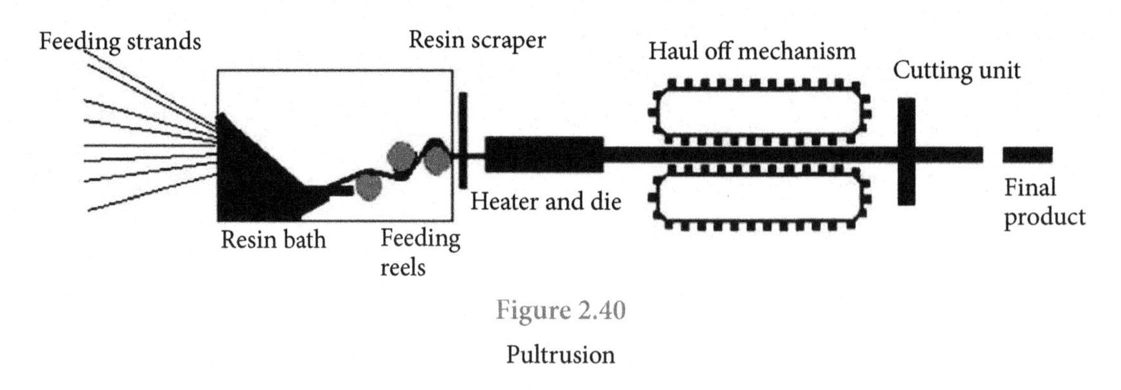

Figure 2.40
Pultrusion

2.1.18 Melt spinning

For the manufacture of synthetic textiles it is necessary to produce polymer fibres. To be useful the fibres must have a relatively high tensile strength and stiffness (depending on the application). For these properties a high degree of crystallinity is required and therefore this process is applicable to crystalline polymers such as PP, Nylon 66 and PET.

There are three types of spinning relevant to synthetic polymer fibres: from the melt and from concentrated polymer solution (either dry or wet spinning), but solution spinning is beyond the scope of this book.

The most important melt spun synthetic fibre commercially is PET polyester. The molten polymer is fed through an extruder and then through a filter into a special die called a spinneret which has a number of fine holes (e.g. 24 to make fibres of approx. 0.5 mm diameter each). Once this extrudate is cooled it can be uniaxially stretched by a godet (take up roller) which increases the fibres crystallinity, and then wound onto bobbins.

2.1.19 Electrospinning

When polymer filaments are created using a electrostatic force the process is called electrospinning. In this process an electrically charged jet of polymer solution or melt is accelerated from a high voltage power supply (kV) attached to an electrode at one end of the solution path towards a collection device which is earthed. Collection devices can range from a simple earthed metal plate or foil through to more complex rotating devices. The main factor is that is it earthed to act as a target for the charged fibres.

The process has a number of requirements:

- A suitable solvent is needed to dissolve the polymer, which must then evaporate quickly within the target plate region.
- Viscosity and surface tension must be suitable.
- Power supply must be adequate to overcome viscosity and surface tension and sustain the jet.
- A suitable gap must exist between polymer solution feed and collection device.

Provided that these criteria are met, electrospinning can produce very fine polymer threads. These can be especially useful for medical products such as artificial scaffolds for tissue growth, and also for fine filtration applications. The renewed interest in biopolymers has also led to many new applications opening up for materials of this type with both plant and animal protein based materials being electrospun. A sample of fibres produced by this technique is shown in **Figure 2.41**, the fibres are so fine they need to be viewed through a scanning electron microscope (SEM) to take pictures of them. These are nano sized fibres, (classified as below 100 nanometres in one dimension), and at least 50 times smaller than the nylon fibres used for ropes for example, which are typically 5 micrometres (5 μm) in thickness.

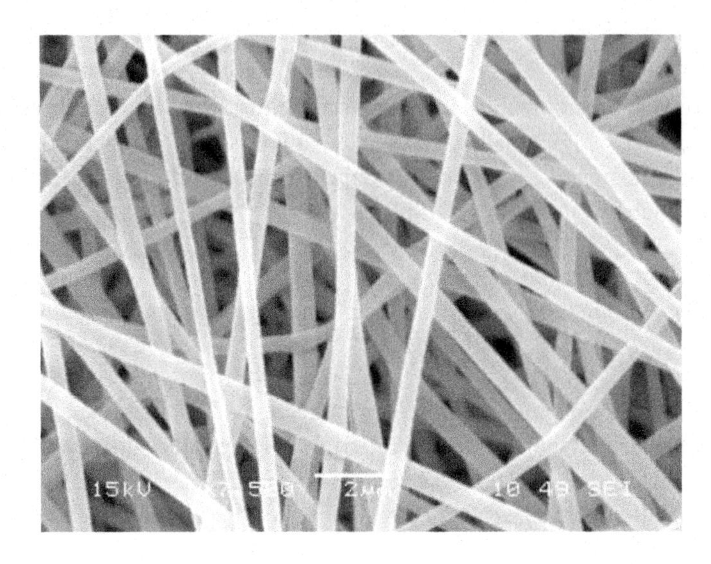

Figure 2.41

Electrospun fibre

2.1.20 Producing plastic foams

A plastic foam consists of a series of cells (hence they are also called cellular plastics) in which gas is trapped and replaces the solid plastic material. By replacing plastic with gas, very low density materials can be produced, for example a polystyrene pot weighing 100 g can be replaced by a foamed pot weighing just 20 g. This is a massive weight saving of 80%, and in terms of raw material usage, it is clear that using foam structures can allow substantial cost savings. However, foaming a material also affects its properties, and the amount depends on the size and shape of the foamed structure. The structures can be characterised by the size and form of the cells, and described by terms such as open cell (the bubbles are loosely connected), closed cell (each bubble is roughly spherical and exists individually) and micro-cellular (very small bubbles, either open or closed). Open and closed structures are simply represented in **Figure 2.42**. Bath sponges are a commonly seen product with a natural open cell structure, whereas a closed cell foam will not absorb water in the same way. Bubble wrap, with all its individual closed cells, is a good visual guide to what a blown up picture of a closed cell structure would look like.

Foaming is commonly incorporated into extrusion and injection moulding and less so with blow moulding, although foamed products can be produced this way. Commonly used plastics for foaming are polyethylene, polypropylene and polystyrene, polyurethane and phenol-formaldehyde. Both thermoplastics and thermosets can be foamed. Foamed sheet products can be used as feedstock for thermoforming foamed cups and fast food trays.

Chemical blowing agents (chemical foaming) or gases (physical foaming) are mixed intimately with the plastic material either in the hopper in the case of blowing agents or by introduction straight into the melt stream if using gases. Blowing agents liberate gas under the heat and pressure of

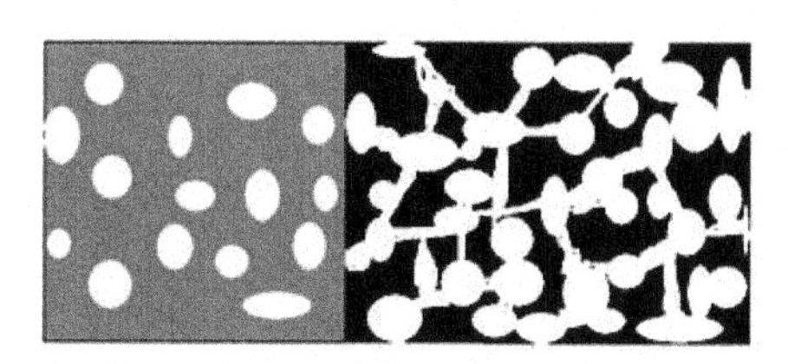

Figure 2.42

Illustration of two different foam cell structures: closed (left) and open (right)

an extruder and the gas is dispersed evenly throughout the resultant product on forming. When the material hardens the bubble structure remains.

2.2 Finishing operations

Finishing operations include assembly and also decorating techniques such as painting and labelling. Assembly and joining techniques for plastics can be split into three major groups: adhesive joining, fusion welding and mechanical methods.

2.2.1 Adhesive joining

Adhesive joining requires the use of another material to create a bonded joint between two materials, which enables transfer of load from one material (adherend) to the other. Unfortunately, thermoplastic materials tend to have poor wetting characteristics, making adhesive joining potentially difficult. (Wetting is the term used to describe how easily an adhesive spreads on a surface. The better the wetting the more intimate the contact, resulting in stronger bonds across a greater area.) Without wetting and therefore contact, bonding cannot occur. Poor wetting characteristics can be overcome with surface modification techniques, however in some circumstances fusion welding techniques may still be preferable.

For joining thermosets, epoxy or acrylic adhesives are used. These have proved to be robust as well as easy to use. It is important when designing adhesive joints between plastic components, that they should be designed to be subjected to shear loads (see Part 3) whenever possible. Exposure to peel loads (effectively pulling apart, like peeling a banana) should be avoided as they are less effective under these conditions.

Adhesive bonds can fail in a number of ways: typically the failure of one adherent, the shear strength failure of the adhesive, or a peel load failure of the adhesive. In fibrous components, delamination of the fibres may also lead to failure.

Other considerations must be the selection of the adhesive for compatibility with the other plastics, the in-service durability of the adhesive material (to include service temperatures, exposures and lifespan) and surface preparation before bonding. Surface modification techniques will be considered briefly later, when we look at decoration.

2.2.2 Fusion welding

Fusion welding uses heat to melt the bonding surfaces of the joint interface. The surfaces to be joined are called the mating surfaces. There are a number of fusion welding techniques, both contact and non-contact, which include hot plate welding, infrared welding, laser welding, hot gas welding and friction welding. The necessary heat energy for melting can be provided by thermal conduction, convection, radiation, friction or induction, and common welding techniques are categorised according to this source of energy in **Figure 2.43.**

Aside from the energy needed to melt the plastic, the pressure on the melted area, which is called the heat affected zone (HAZ), is also very important. This force causes the molten polymer to flow, which allows mixing of the plastic from both sides of the bond to occur. Therefore, once the bond cools and fuses an inseparable layer has been formed. The heating time required to make the plastic melt and flow is therefore also an important parameter.

Whatever the welding type it can generally be achieved in five steps:

- Clean the surface
- Heat the surface
- Apply pressure
- Cool under pressure
- 'Finish' the welding seam if necessary

To weld together two surfaces they must both be molten, therefore the temperature at which this is achieved must be similar for both sides of the weld. The two components must also have a similar viscosity so that they will flow and mix, and for this to occur they must also mix together readily at a molecular level. If two incompatible

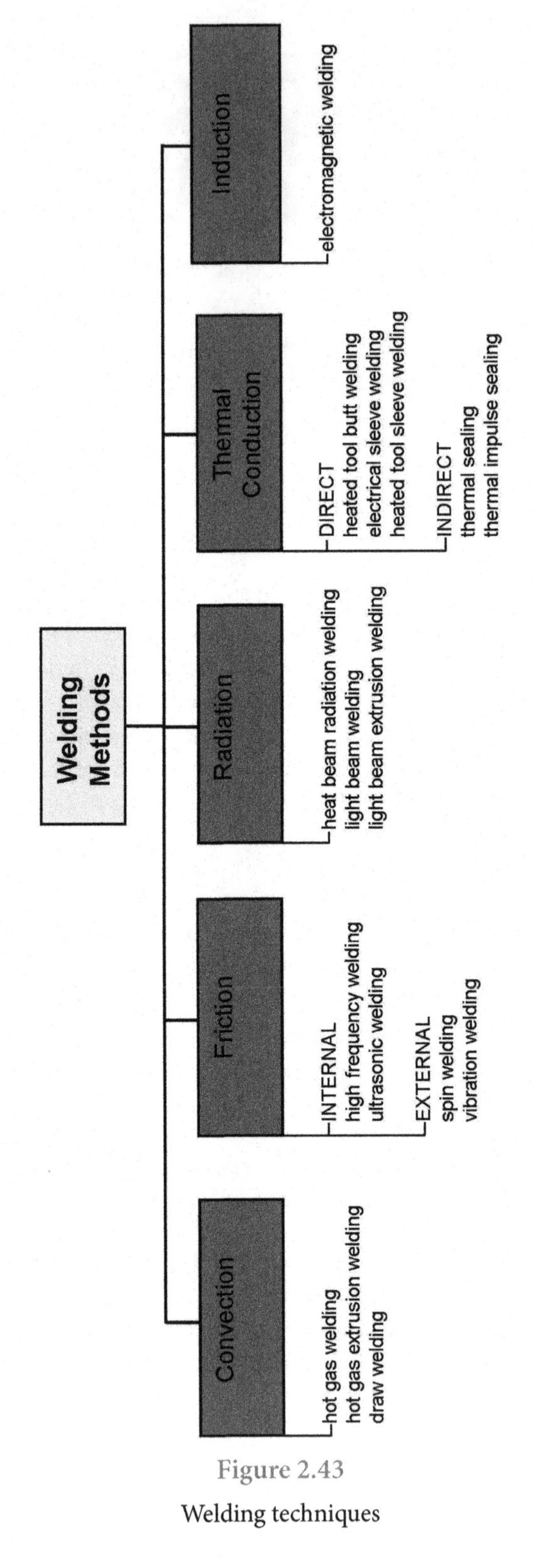

Figure 2.43

Welding techniques

plastics (analogous to oil and water) are used then they will not form a bond. Generally, both sides of a weld are of the same material type.

2.2.2.1 Hot tool welding

In all the variations to this method, heating elements are used to supply heat to the joint. This can be done directly by placing heat at the mated surfaces or indirectly by relying on thermal conductivity and heating the rest of the part through to the surface. The poor thermal conductivity of plastics means that only thin walled parts are joined by indirect methods such as thermal impulse sealing.

Hot-tool butt welding on the other hand is a common method used on PP and PE tubes for example. The join is heated by a heater placed between both surfaces, and then pressure (approx. 22 psi) is applied until a melting bead of polymer material is visible. The heater is withdrawn and a lower pressure used to clamp the joint together. The joint then cools, forming a weld between both sides.

2.2.2.2 Radiation welding

In radiation welding, a source of energy such as light or heat is used. Laser welding techniques also fall into this category. Transparent materials cannot be joined in this way as they transmit the light rather than absorb it.

2.2.2.3 Induction welding

In this case, a third material is placed between the two layers to be mated – rather like an adhesive would be. However, in this case the material is a powder that can be magnetically activated to heat up and supply heat to the surfaces. They are then joined under pressure.

2.2.2.4 Friction welding

This encompasses both ultrasonic and spin welding. The basic principle of ultrasonic welding is vibration. By vibration of one component in contact with another, the heat generated by the movement melts the materials at the interface. A typical ultrasonic welding operation will consist of an ultrasonic generator, a converter, a booster and a metal bar that is designed to resonate at the required frequency. This is called a horn and must be applied as close to the joint as possible. Therefore a supporting fixture is required to hold the parts together. There are a number of variables with this process including materials, part geometry and wall thickness, which all affect mechanical energy transmission.

Spin welding is used for joining circular cross sections. Again as with ultrasonic welding the mechanical force is applied to one half of the join as the other is stationary. In this case, one component is spun to create friction. Once a melt film is formed the rotation is stopped and the weld can form. Heat generation results from both external friction between the component parts and internal friction caused by melt shearing.

2.2.3 Mechanical fasteners

Mechanical fastenings systems are essentially the screw, the rivet and the bolt.

If adhesives or melt fusion techniques are impractical, mechanical fastenings can be used. This could be for cost reasons, if parts may need to be removed during service for repair or replacement, or if assembly is to be done on site, then mechanical fastenings may be preferred.

Bonded joints are generally more efficient than mechanical joints, however there are a number of advantages in the use of mechanical joints including quality control, no surface preparation, and damaged or badly assembled components can be easily dismantled. The assembled joint can be used immediately and even in poor conditions can usually still be assembled if needed.

However, mechanical joints are generally weaker than a bonded joint due to the need to make holes in the component, where drilling may weaken or damage the part. There is also a weight penalty attached to the use of metal fasteners and the cost of the joint is relative to the size in these cases. There can also be problems if used in varying temperature conditions, as a thermal expansion mismatch between plastic materials and metal fasteners can cause a weakening of the assembly.

It is possible to incorporate metal assembly inserts into some shaping processes such as injection moulding. By incorporating an insert as the part is formed in the tool, in a process termed as over-moulding, the need for drilling the part after manufacture is avoided. Other mechanical assembly devices such as press fitting (see next section) can also be incorporated in the shaping stage.

Using the over-moulding technique in injection moulding can also be an assembly method in itself. Consider a screwdriver for instance. This is a metal driver with a plastic handle. By placing the metal driver in the tool and moulding the handle straight onto it, the need for further assembly operations is removed. The plastic shrinks onto the metal creating a very strong mechanical bond. This is commonly done using a robot to accurately place the metal component in the tool before each moulding cycle commences.

2.2.4 Mechanical joints

2.2.4.1 Press fitting

This is a very simple assembly method where one part is forced into a mating component. This has the advantage that no additional components such as fasteners and no further processing are required. Since press fittings can be produced to join either the same or different materials, joint design is an important factor. Generally, the main consideration would be creep properties, which relate to a material's life cycle. Examples of commonly used press fits include the lids on Tupperware and the fit between margarine tubs and lids. In an example of a permanent press fit, plastic components such as bearings and wheels are often fitted onto connection shafts, including plastic to metal press fit joints. This kind of joint can however induce stress in the plastic part which must be considered in their design.

2.2.4.2 Snap fitting

Like a press fitting, a snap fitting requires no addition components or joining techniques. A snap fit consists of a lock and groove design. The hook is deflected during assembly but returns to the original position once inside the mating component. The hook and groove configuration gives a good gripping force. Again, this design can be used between both similar and dissimilar materials. Plastic ties commonly used in gardening are good examples of this design, and are incredibly hard to reopen once in use.

2.2.4.3 Living hinges

A living hinge is one that is integral to the component, allowing two parts of the same component to repeatedly open and close (or flex). The most commonly used materials are PP and PE. This is partly due to their low cost and easy manufacturing, but the main property they possess is the ability to flex over and over again without material failure. This will be further considered in Part 3. An example of a commonly seen living hinge design on bottle lids is shown in **Figure 2.44**.

2.2.5 Decoration: painting, plating and printing

2.2.5.1 Self coloured plastics

The easiest way to decorate a part is to use a self coloured plastic. Additives are used to provide the colour in the form of pigments or dyes. Pigments work by evenly dispersing throughout the polymer matrix, dyes work slightly differently by dissolving right into the plastic.

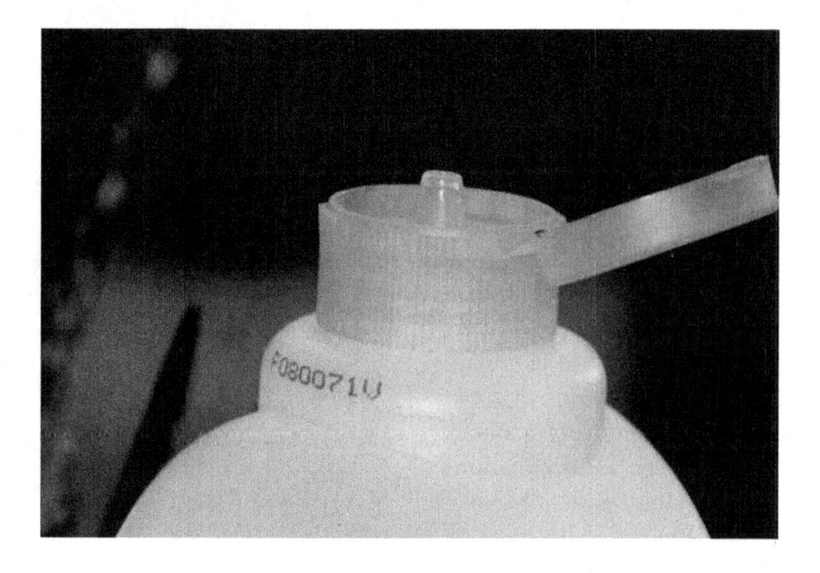

Figure 2.44

Living hinge on a bottle lid

Special effects pigments can be used to provide surfaces that look like metal or are pearlescent, change colour with temperature (thermochromic pigments) or change colour depending on the angle you look at them from (interference pigments). This allows plastics to come in a variety of design effects which conventional materials such as metal cannot hope to match without requiring finishing steps such as painting.

For common and relatively transparent materials such as polypropylene, polystyrene and polyethylene there are literally thousands of colour variations available. For plastics with a strong base colour or those that require very high processing temperatures, the possibilities are more restricted, as the range of available colourants is significantly reduced as processing temperatures rise.

Pigments can be added in at the compounding stage or at the processing stage. However, as handling pigments can be very messy, colour concentrates pre-dispersed in plastic granules (called masterbatches) are generally used in non-extrusion applications such as injection moulding, blow moulding or film blowing operations. Masterbatches can be added to uncoloured plastic resin at 1-5% (usual usage range) and during plastication they become mixed with the resin to provide an even colour.

Pigmentation remains the most common method of decoration for plastics in the marketplace today and it is a cheap way of providing colour products. The disadvantage is that the component will be all the same colour. Sometimes, this can be a drawback in consumer goods, and as an alternative to this, manufacturers may utilise different coloured parts in an assembly of components to add variety in the way their products look. The example of the lawnmover in **Figure 2.45** uses four different colours on the injection moulded components, with the cable matching a button on the handle assembly.

Figure 2.45

Lawnmower

Source N. Goodship

2.2.5.2 Painting plastic parts

In the lawn mower assembly in **Figure 2.45**, it was not necessary to accurately match the colours of components made from different materials. However, if considering the exterior of a car, there are not only plastic components to match, but also metal ones. Where accurate colour matching like this is required, especially when using lots of different base materials, painting is often the preferred decorating technique. Painting is commonly used in both automotive and consumer electronic applications.

Painting is the application of a liquid, which is then dried and solidified. Because the paint must stick to the surface of the plastic, it is necessary to ensure that the plastic has a suitable surface free from dirt and grease. For some plastics, like polypropylene and high density polyethylene, the surface cannot be painted without pre-treatment techniques. The paint would literally just fall back off again and not adhere. To activate the surface a number of pre-treatments can be used such as flame treatment, plasma or corona discharge. This pre-treatment may be manual or automated, depending on the size, shape, number and complexity of the plastic components. A bottle being manually pre-treated is shown in **Figure 2.46**. This changes the properties of the surface and allows the paint to stick. Primers can also be used. A primer is an additional layer between the substrate (plastic) and the paint which allows adhesion to occur.

They are various types of materials used for paint formulations; some based on solvents and some based on water. Solvent based materials give off volatile organic components (VOCs) as they evaporate and dry. The amount of VOC that can be released in the environment was limited in the UK by the Environmental Protection Act (1990), and this has led to an increase in the usage of water based paints. These tend to be acrylic based and easy to apply using water instead of solvent. However, compared with solvent based paints they usually need more heat and time in order to dry.

Like plastic formulations paints have 'recipes' and contain varying amounts of solvents, binders, resins, plasticisers, pigments and or dyes and extenders, depending on the formulation required.

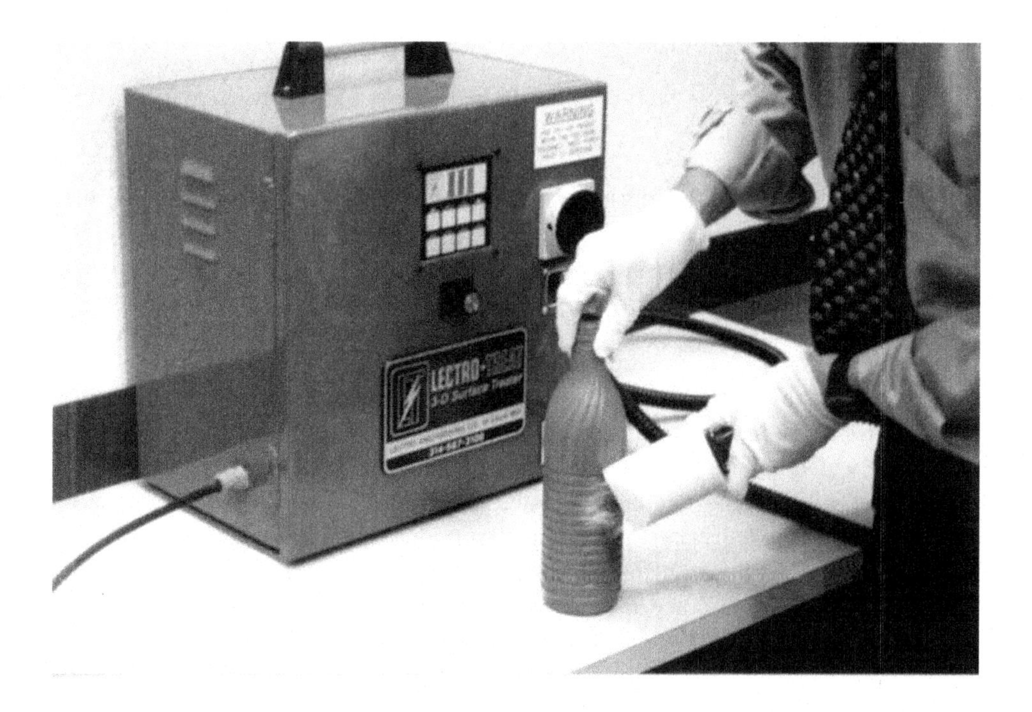

Figure 2.46

Plasma surface treatment of an HDPE bottle

A basic painting process consists of the following steps:

- **Part preparation**: A pre-treatment or priming processes.

- **Paint application**: There are various methods e.g. air atomisation, airless atomisation, electrostatic assist, brushing, wiping, pouring, trowelling, flow coating and centrifugal coating. Further discussion of painting methods is beyond the scope of this book.

- **Drying**: The process required to turn a wet paint layer into a dry film can be physical or chemical. In a physical drying process, as the solvent evaporates the remaining material will simply stick to the surface. Chemical drying is equivalent to the crosslinking of thermoset plastics. Irreversible chemical bonds form as the paint cures. Like thermoset plastics, it is also common to use heat to bring about this reaction and large ovens are usually used to cure parts. Radiation can also be used to initiate cure and can be in the form of ultraviolet, microwave, infrared, electron beam or plasma.

Automation on paint lines is common especially in the automotive industry. Robots are used to spray paints, and it is in high volume manufacturing that the advantages of spray painting are most applicable. However, set up costs and energy usage in paint shops are both very high, and with increasing environmental legislation on both emissions and energy use, many manufacturers are turning to alternative decoration techniques for their products.

2.2.5.3 Metal coating

The application of a metal coating onto plastics can be done by two methods: plating and vacuum coating.

Vacuum coating applies the metal as a vapour or as relatively small particles. The main metal used here is aluminum and this is applied to products such as trophies, bottle caps and metallised films and foils. The films can be used in subsequent decoration techniques such as in-mould decoration.

Plating involves the deposition of a thin layer of metal by electrolytic or electroless means.

Electroless plating is the less frequently used technique, however it is used as a pre-process for subsequent electroplating in the production of printed circuit boards for instance. Electroplating is also used more widely to give plastics a metallic appearance. Not all plastics are suitable for this process; ABS is the material most commonly treated in this way. It is widely used in the car industry to produce chrome-like and polishable surfaces on plastic components. Electroplating uses an electric current to lay metal onto a conductor (the plastic). Therefore the plastic must first be made conductive to undergo this process. This is either done using electroless plating first, or by adding a conductive additive to the plastic.

2.2.5.4 Dye sublimation

This process is also known as thermal transfer or dye diffusion.

Dyes are commonly used as additives to colour plastics. In dye sublimation they penetrate only into the plastic surface and sit just below the surface layer. A pattern is first printed onto a paper or plastic sheet using special sublimation printing dyes. When brought into contact with a plastic under heat, the dye vapourises, turning from a solid directly to a gas (this is sublimation). The gas formed can then diffuse into the plastic surface from the sheet, where it then returns to its original solid state. Special dyes suitable for this process are required, but the designs have good abrasion resistance as the dye is below the surface of the component. A wide range of decorative effects are possible with this technique, which is suitable for products such as computers and mobile phones (**Figure 2.47**). A similar process is used for iron-on T-shirt transfers, and in fact, it is from the textile industry that this process originated.

Figure 2.47

Metallic finish mobile phone cases subsequently decorated by dye sublimation (top)

2.2.5.5 Hot stamping

Hot stamping uses heat and pressure to transfer a foil from a carrier film to a plastic substrate. The film material or foil is typically multi-layered:

- Layer 1 (top): a thin film carrier usually made up of polyester (although polycarbonate and cellophane can also be used).
- Layer 2: a release layer. This holds and protects the print layer but only until heat and pressure are applied.
- Layer 3: a decorative coating (any colour, often a metallic effect).

- Layer 4: an adhesive coating especially tailored to adhere to the type of plastic substrate. The application of heat and pressure (for a short time only, typically one second) activates this layer and melts the plastic layer on the substrate producing a strong bond. The heat also activates the release layer, causing the carrier layer to separate from the finished component.

An elastomeric or metal die of the required pattern is heated and used to stamp the foil. These typically operate in an up and down motion driven by pneumatics or hydraulics. The die is fixed to a heated platen. Feeders are used to refresh the foil after each stamp and enable fast cycle times. Hot stamping is generally used for single colour and low resolution printing. One downside of this process is the cost of the foil feedstock which can be expensive.

2.2.5.6 Printing

Like painting, printing is the application of a liquid coating but in this case only over a small area. Paint or printing inks can be used, depending on the technique. Typical printing techniques include screen printing, flexographic printing, gravure printing and pad printing.

Screen printing: This is a simple and inexpensive method and suitable for both large and small areas. However it is slow, only suitable for short runs, and only flat or cylindrical surfaces can be decorated. It is a technique used in the textile industry.

The screen itself consists of a stencil of the desired pattern which is bonded to a fabric and tensioned in a frame. Ink is pushed through the screen by an applicator and onto the substrate. The inks then require drying or cure by UV radiation. Hot melt inks which then solidify can also be used. These machines can be automated or manual.

Flexographic printing: In this method the ink is carried on the raised portions of an engraved plate which is transferred to a film when pressed onto it. Plates can be made most cheaply from elastomeric materials or from more expensive photopolymers for greater definition. This process does not use metal printing plates. The technique is typically used in flexible packaging applications.

Gravure printing: This is the transfer of ink from recesses within a rotating cylinder, and like flexographic printing it is used in flexible (film) packaging applications. This technique is however more expensive and not suitable for bottles, which can be decorated using the flexographic technique. The gravure process is applicable to many plastics in sheet and film form and good resolution on tones and shading can be produced. It is capable of producing patterns such as wood grain on films which can then be used in other techniques such as hot stamping, which has already been described.

Pad printing (also known as tampo printing): A number of materials can be printed by this process including thermoplastics, thermosets and natural materials such as wood and leather. Some materials may require pre-treatment but a wide variety of colours can be achieved with this process. A further advantage is that multiple colours can be applied during one process cycle and high registration accuracy is achieved. Specialist inks are used which consist of a solvent, resin, binder, pigments and fillers. Pad printing is similar to the gravure technique but suitable for small and irregular shaped objects. Ink is first applied onto an engraved plate called a cliché. Then, once the excess is removed, this ink is picked up by a silicone pad which is used to print onto the substrate. The use of soft silicone allows recesses and irregular surfaces to be printed. It is a relatively simple and inexpensive technique used to achieve fine lines on small areas and uses similar inks to those used in screen printing. It can be applied to components such as cigarette lighters, pens and spectacle frames.

2.2.5.7 Laser marking

Laser marking is a non-contact process. Like laser welding, the laser causes the plastic to melt. It can also be used to vapourise a surface or discolour it, thereby allowing marks to be distinguished. If laser marking onto laminate materials, it is possible to expose the material below the surface. This could consist of a layer of a different colour, or be a different material entirely such as a metal.

2.2.5.8 Labels

Fixing labels to the surface of a component requires a pre-cut and pre-printed product. An adhesive is required to attach it, and this can be pressure sensitive or heat activated. Adhesive labels on packaging must be easy to remove with water to aid recycling.

2.2.5.9 In-mould decoration

Post-processing operations such as component painting, plating and printing all add cost to the manufacture of products. Inserting labels or decorative decals during processing is now seen as an economic way to decorate plastic components, as this removes the need for post-moulding processes. This can occur at varying times during the processing operation depending on the form and function required.

In injection moulding, blow moulding and compression moulding processes, an insert can be placed in the mould tool prior to beginning a moulding cycle (see **Figures 2.48** and **2.49**). A variety of materials can be inserted in this way depending on the individual process, from thin labels and films to real wood veneer and textiles. All remove the need for further finishing, as a complete decorated article is produced by the machine. This is effectively the same as the over-moulding process described in section 2.2.3 for incorporating metal fixings or assembling a screwdriver.

In the packaging industry, where vast quantities are items are produced, in-mould labelling is a very common manufacturing technique not just in flat surfaces but up to eight sided wrap around labels. It is estimated over 85% of food packaging production utilises in-mould labelling. These are commonly seen on items such as yogurts pots, deserts cups and butter containers (injection moulding applications). The sleeves on detergent bottles are a routine example of in-mould labelling of blow moulded products.

Thermosets as well as thermoplastics can benefit from in-mould decoration techniques. Overlay moulding is commonly used particularly with melamine-formaldehyde compounds. Items such as dinnerware which are compression moulded can have a very thin decorative melamine-impregnated paper placed in the mould tool cavity prior to final cure. The mould tool is opened to allow for this step.

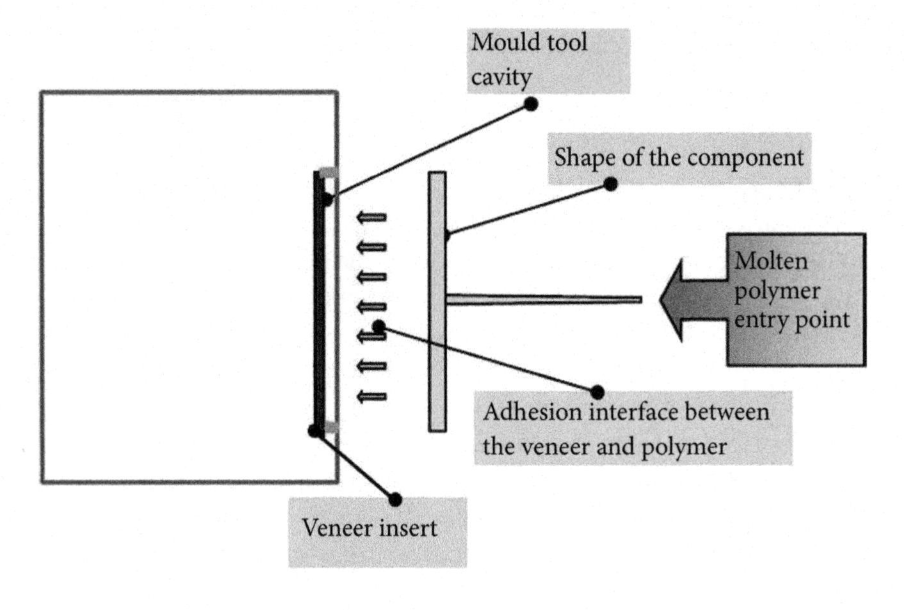

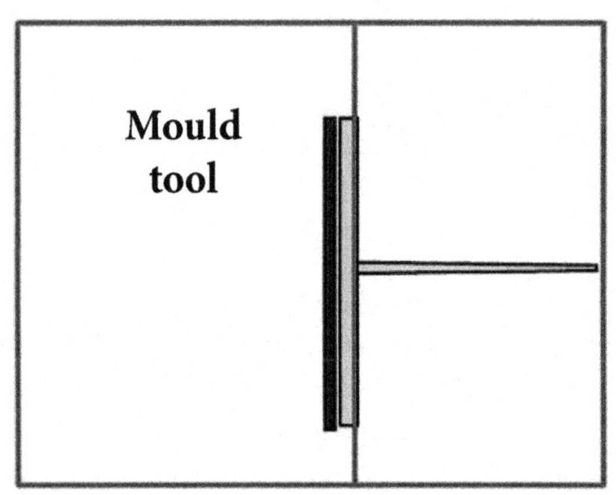

Figure 2.48

Principle of insert moulding with injection moulding

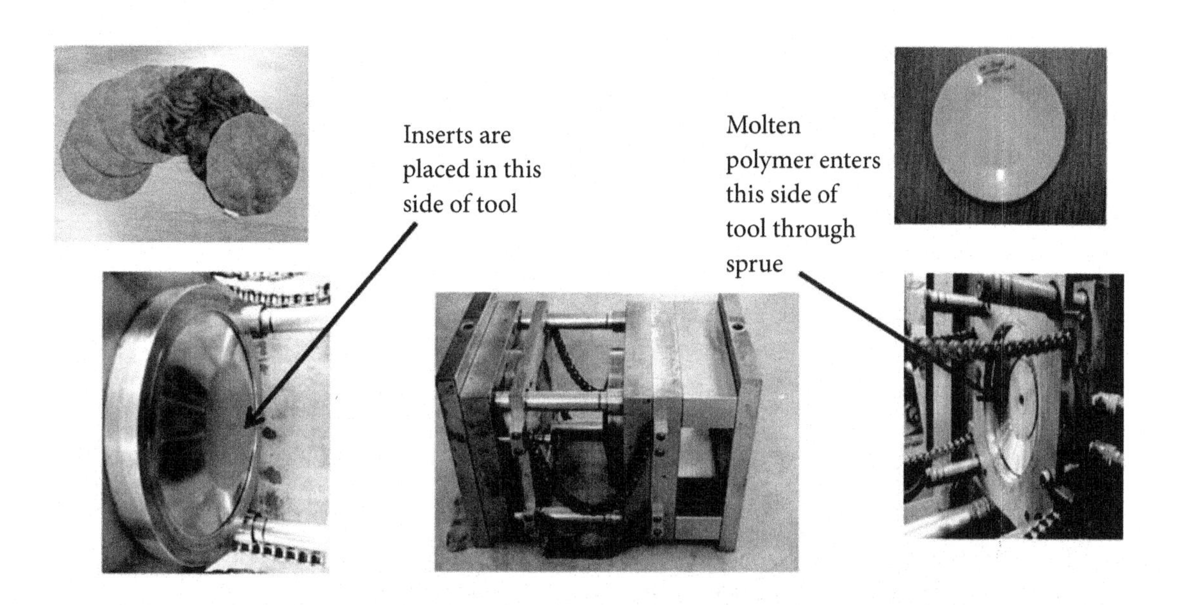

Figure 2.49

Practical example of insert moulding with injection moulding

Part 3. Properties

3.1 Quality and testing

The plastics business is continually evolving as new grades of plastics and additives are regularly introduced. (Entirely new materials appear in the marketplace much less frequently.) At the same time, new machinery and variations on established processes are also offered, and customer demands alter in response to market and other pressures. Cost reduction, greater yields, better products, higher safety and more environmentally friendly performance are all driving forces behind quality improvements and improved efficiency. The capabilities of plastics to meet new and demanding challenges increase all the time, but this also results in a potentially bewildering choice of materials and processes available to the designer and specifier, together with the possibility of poor selections and product failures.

Specifications for the properties of plastic materials are clearly essential. They not only allow comparison of material properties for designers but also enable monitoring of quality control within and between companies and customers. There are many national and international standards, and some of these will be mentioned in the following pages. However, as well as these standards, quality and performance requirements may also be defined by company-specific (in-house) quality measures. For example in the automotive industry, manufacturers often have their own individual requirements, whether for the plastics in the dashboard or the final standard of the paint surface finish. Before these specifications are applied, it is necessary that there is some point of agreement that these attributes are important to quality. This may involve the material supplier, the plastics processor and the customer to ensure consistency in supply from batch to batch.

In some cases a standard set of tests may be needed that can be applied consistently at many different sites by different operators. An attainable and repeatable means to achieve them therefore needs to be agreed and specified. Quality factors like test methods must therefore look at a measurable aspect of quality that can be monitored accurately. For in-house production it must also be controllable.

A further aspect of quality and quality improvement is documentation. It is necessary to keep records to allow examination. This would allow an investigation of process fluctuations for instance, or the identification and elimination of unreliable suppliers. Quality is continually subject to examination for the potential for improvement.

For a company that compounds plastic, and also for the raw material producers, material quality considerations must be applied to all raw materials being used. This includes the base polymer and all the various additives that may be included in a formulation. A processor would monitor and record processing parameters for all materials, as well as conversion rates, outputs etc., and also the quality of the product it produces. To do this it is necessary to carry out testing.

With quality, responsibility, liability, health and safety all major factors in today's marketplace, it is paramount that producers are aware of the safety and reliability of their products. They must also be able to provide that information to their customers. This type of detailed performance information can also be beneficial in marketing new products to pre-identified targeted applications. It allows potential customers to quickly ascertain the performance of new materials against current materials. If there are problems with materials once they are in use such data can aid in providing customer support for quick troubleshooting of performance problems.

3.2 Introduction to common methods

Test methods and procedures for testing of plastic and polymer materials are constantly being developed and reviewed. There are literally hundreds of standards relating to the performance of plastics. There are international bodies responsible for developing standards such as the International Organization for Standardization (ISO), and the European Committee for Standardization (CEN). There are also national standards bodies whose methods are widely used, such as ASTM International, (originally known as the American Society for Testing and Materials – ASTM). Other commonly cited national standards organisations include the British Standards (BS), Japanese Industrial Standards (JIS) and the German Institute for Standardization (DIN).

This book cannot hope to cover all tests but will outline some of the most common.

However when the physical properties of a polymer or plastic material are quoted, what do they actually mean?

For example, properties commonly reported by materials' suppliers include tensile, flexural, compression, impact and creep data. These are **mechanical properties.** These values are amongst the most important criteria for selection purposes as they provide information about material loading tolerances (such as how much pushing, pulling, crushing or dropping force a material can take) and the likely responses to those forces. This allows a designer to decide if a material can withstand the conditions which will be encountered in service without breaking.

For example:

- Will a plastic shopping bag split if an extra tin is put inside? (Tensile behaviour)
- Will a plastic drinks cup buckle when gripped? (Flexural)
- Will a plastic automotive lens crack when a stone hits it? (Impact)
- Will a mobile phone case split if I drop the phone? (Impact)

- Will a load-bearing component change shape or rupture after being in service and under load for five years? (Creep)

However, it is first worth considering how these properties compare with other materials such as wood, ceramics and metals. Generally the properties of plastics improve as they become more expensive (no surprise there). The cheaper commodity materials offer lower properties and performance than those described as engineering materials. Speciality polymers tend to have at least one property which is particularly remarkable, such as improved heat resistance, non-adhesive properties etc. For super-strong materials with comparable properties to some metals, it is necessary to look at reinforced composite materials, however, the strongest materials available to designers for large-scale use are still metals.

In some applications, plastics can replace metal components and their lower density compared to metals is often advantageous for product weight. Like plastics, there is a range of metals available, some with comparable properties to plastics. For weaker materials like wood and glass, polymers often fall within or exceed the property requirements of these materials. A glass bottle for instance, while it makes an excellent container, cannot be squeezed like a plastic one. A 'squeezy' bottle therefore has to be made of plastic. This discovery helped launch the blow moulding industry – see Part 2. A plastic bottle is also much lighter than a glass one because glass is more dense than plastic.

Plastics can therefore be thought of as having a low density, intermediate strength (compared to metal, wood and glass) and intermediate stiffness. However within this 'intermediate' range, a vast array of tailored properties can be achieved.

Other plastic properties which may be considered include those important for manufacturing processes such as heat stability, rheology (deformation and flow), abrasion resistance, melting temperature or softening point, shrinkage and solidifying temperature. Further properties may relate to outdoor performance and resistance to daylight, rain and ultraviolet (UV) radiation.

An overview of all the properties included in this section is given in **Table 3.1** along with the relevant section number in the book, and an example of an International Standard (or alternative). More details of each test can be found in the relevant text. Properties covered in this part of the book are those listed in most plastic data sheets supplied by polymer and plastic manufacturers. More generalised details of some other useful properties beyond this general scope can be found in section 3.7.

Table 3.1
Common plastics properties, tests and units

Property	Section number	Test standard no.	Units quoted
Physical properties			
Density	3.3.1	ISO 1183-1:2004	g/cm^3
Water absorption	3.3.2	ISO 62:2008	%
Humidity absorption	3.3.2	ISO 62:2008	%
Melt flow rate	3.3.3	ISO 1133:2005	cm^3/10 min
Viscosity number	3.3.4	ISO 307: 2007	mL/g
Mechanical properties			
Tensile	3.4.1	ISO 527-2:1993	MPa
Flexural	3.4.2	ISO 178:2001	MPa
Compression	3.4.3	ISO 604:2002	MPa
Impact, Charpy	3.4.4	ISO 179-1:2000 and 179-2: 1997	kJ/m^2
Impact, Izod	3.4.4	ISO 180:2000	kJ/m^2
Falling dart impact	3.4.4	ASTM D5420 - 04	J
Creep	3.4.5	ISO 899-1:2003	MPa
Hardness (Shore hardness)	3.4.6	ISO 868:2003	Scale (A or D) /time (seconds):hardness value (no units) e.g A/15:55
Thermal properties			
Melting point and other transitions	3.5.1	ISO 11357 (several parts)	°C
Heat deflection temperature (HDT)	3.5.2	ISO 75 – 1:2004	°C
Vicat softening	3.5.3	ISO 306:2004	°C
Thermal expansion	3.5.1 & 3.5.4	ISO 11359-2:1999	°C^{-1}
Thermal conductivity	3.5.5	ISO 22007-1:2009.	W/mK
Flammability	3.5.6	UL 94	H or V with rating scale (V-0 to V-2)
Electrical properties			
Volume resistivity	3.6.2	IEC 60093:1980	Ωm (Ohm meter)
Surface resistivity	3.6.2	IEC 60093:1980	Ω (Ohm)
Tracking index	3.6.2	IEC 60112:2003	V (Volts)
Electrical strength	3.6.2	IEC 60243-1:1998	kV/mm

3.3 Physical properties

3.3.1 Density

The densities of plastics are very low compared to those of metals as shown in **Table 3.2**, and some plastics like polyethylene (PE) and polypropylene (PP) are actually lighter than water and will float. Most plastic materials have a density between 1.0 and 1.5 g/cm^3. Polytetrafluoroethylene (PTFE) at 2.2 g/cm^3 is an example of a very dense plastic material.

Table 3.2
Density of common materials

Material	Density (g/cm^3)
Wood	0.2-0.95
Water	1.0
Plastics	0.9-2.2
Low density polyethylene (LDPE)	0.9
Poly(tetrafluoroethylene) (PTFE)	2.2
Aluminium	2.7
Steel	7.8

The importance of density to designers is particularly illustrated by the drive to reduce the weight of cars and thereby reduce fuel consumption. Replacing a dense steel component with a plastic one can bring about massive weight savings and therefore better fuel economy and lower running costs.

The importance of density to processors is more in terms of how much bulk space is required for storage and transport. For example how much material will fit in a hopper, or a machine screw during processing, or what weight of finished components can be fitted into a container. The light weight of plastics can have drawbacks, and the logistics of shifting plastics become especially apparent in their recycling. A lorry load of used plastic bottles will weigh considerably less than a lorry load of aluminum cans. The cost implications of this are considerable for the waste recycling infrastructure where plastics make up a lot of the volume, but not much weight in the household waste stream. Therefore, the density of a material is an important consideration for processing, performance and disposal.

As with many of the properties of plastics, it is possible to alter the density to make a material lighter or heavier. Foaming can be used to introduce air and produce a lightweight component, or at the other extreme, very heavy fillers can be used to make the plastic denser. However, density is rarely adjusted for its own sake, but a change in density is often a side effect of adding other functionality.

3.3.2 Water and humidity absorption

Immersion in water can cause swelling in some plastics, resulting in dimensional changes in plastic parts. If there are water soluble additives or components these can be dissolved into the water (and there are even some water soluble polymers). In some cases, the properties of a component can vary as they are dependent on the level of moisture, especially with hygroscopic (water-absorbing) materials such as nylon. That is why plastic components such as nylon have specified conditioning regimes and 'before and after' conditioning data are usually presented to ensure compatibility of results. **Table 3.3** gives some water absorption values for a number of plastic materials.

Table 3.3
Water absorption of various plastic materials

Material	Water absorption (% in 24 hr)
Nylon 6 (PA 6)	1.9
Nylon 66 (PA 66)	1.5
Polyacetal (POM)	0.25
Poly(tetrafluoroethylene) (PTFE)	0
Polypropylene (PP)	0.02
Poly(vinyl alcohol) (PVAL)	Dissolves

In ISO 62 the water absorption is measured at 23°C, in boiling water and at 50% humidity for a prescribed number of hours, and the weight of the sample before and after is taken. These values can be expressed as percentages. It is also possible to calculate the diffusion coefficient and the water content at saturation using this standard. Diffusion coefficient relates to water barrier properties, and high or low values may be desirable for specific applications.

3.3.3 Rheology

In Part 1 it was shown that the molecular make-up of polymer chains (both length and structure) can affect a number of material properties. However the flow behaviour of a plastic is also dependent on all the other materials that it contains, such as the types and amounts of fillers or other additives, as well as the polymer itself.

The ability of a plastic material to flow often limits its use in certain plastic processing operations. For example in Part 2, it was shown that the flow requirements are different for injection moulding (fast flowing and therefore low viscosity materials) and extrusion blow moulding (stiff, high melt strength, high viscosity materials). Plastics are viscoelastic having both elastic and viscous properties. In the molten state it is the viscous properties that are of most interest, therefore viscosity is used to define the flow under the main processing control parameters – shear levels and temperature.

As stated viscosity can be considered dependent on material properties such as chemical structure, chain length, branching levels and additives, and also process properties including temperature, pressure and shear rate.

At the high shear rates experienced in processes such as extrusion and injection moulding, polymers exhibit a behaviour called shear thinning. This means that at higher shear rates the viscosity of the material decreases, and therefore it becomes more free-flowing as shown in **Figure 3.1**.

The most common measure of flow used in the plastics industry, especially in quality control operations, tends to be the Melt Flow Index. This is known by a number of abbreviations such as MI, MFI or MFR, but they are all essentially the same test.

In this procedure, a machine measures the rate of extrusion, that is the amount of material exiting an orifice of a known size, diameter and length, which is pressured by a prescribed load (as defined in standards, example weights are 2.16 kg and 5 kg) and temperature. Only one level of shear is exerted throughout this test and this is relatively low compared to levels experienced in injection moulding. In other more advanced techniques, a range of shear levels can be used.

The measure of MFI is the reverse of viscosity. Therefore, a material of high viscosity (such as treacle) would have a low MFI, and a material which is very free flowing i.e. similar to water, would have a very high MFI but a low viscosity. This can sometimes cause considerable confusion.

Consider a common material such as polypropylene which is tested at 230°C with a 2.16 kg load (ASTM D1238, ISO 1133). Processes such as blow moulding require a high viscosity, and therefore fractional MFI values <1.0 cm^3/10 min, whereas injection moulding may use materials in a range from 2.0-50 cm^3/10 min. The units relate to the volume extruded over a ten minute phase.

As MFI is influenced by the load on the material and the testing temperature, material data tested under different weight regimes cannot be compared. Because this method uses a low shear rate to test, the effects of shear thinning are to some degree removed and therefore for a polymer material this method can provide an indication of molecular weight.

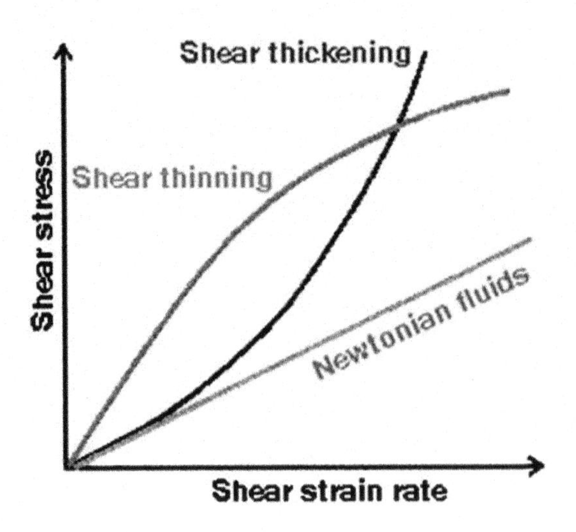

Figure 3.1

Variations in fluid behaviour under shear stress, showing shear thinning

3.3.3.1 The capillary rheometer

As we saw in Part 2, polymer processing operations are conducted at a variety of temperatures and shear rates, and both of these factors influence shear behaviour. For this reason the ability to design viscosity experiments which can vary conditions of temperature, shear and dwell is extremely useful, especially for polymer processing machine designers.

For these kinds of experiments, a capillary rheometer can be used. In essence the testing procedure (ASTM D835) is very similar to that for MFI, the material is extruded by forcing it out of a known sized die by exerting a known force. However in this case the force can be varied, and material behaviour at a number of shear rates can be obtained from one test. In this way a graph can be generated showing the behaviour of the plastic or polymer at different shears. This is a curve, because as stated earlier, polymers are shear thinning materials.

3.3.4 Viscosity number

The viscosity data discussed so far relates to the behaviour of the polymer when molten. Viscosity can also be measured in solution, and the values produced by this method relate to the molecular mass of the polymer. However if other fillers and additives are included in the formulation the result may not be directly correlated as their presence may change the solution properties. In these cases the method can be useful for quality control purposes rather than molecular mass determinations. ISO 307 is such a method used for polyamides (nylon). This material can be produced in a variety of molecular configurations e.g. PA 6, PA 66, PA 11, PA 12, PA 15, PA 69, PA 610 are just some of the possible types.

A known amount of the nylon is dissolved in a known amount of solution of a specified concentration. Both the pure solvent and the nylon-containing solution are then timed travelling through a fixed volume of a viscometer which has been placed in a temperature controlled water bath. Using the nylon concentration in the solution and the relative flow times, a viscosity number (VN) can be calculated, the units are mL/g.

3.4 Mechanical properties of plastics and their effect on performance

As already discussed, mechanical properties are used to describe how a material will behave under load. This can be presented by properties such as strength, hardness, toughness, elasticity, brittleness and ductility, and these broadly describe the levels and types of force or stress that the materials can withstand.

Properties such as 'strength' can have several meanings. For example tensile strength, flexural strength and compression strength relate to the ability to withstand pulling apart, bending or compressing respectively. So within the context of mechanical testing, it is important to be more specific in discussing the type of force applied rather than simply saying that a material is 'strong'. Common types of stress a plastic can experience are tension, torsion, compression, impact and shear, or perhaps a combination of these stresses as might be experienced during fatigue testing. It is therefore also worth defining what is meant by other terms related to mechanical properties:

Brittleness and Plasticity: These two properties are the opposite of each other. One is a deformation without breaking (plasticity) and the other breaking (shattering) before deformation (brittle behaviour). Glass is a brittle material, as are a number of clear amorphous plastics such as polystyrene. Semi-crystalline polymers such as PP, LDPE and nylon exhibit plasticity and can be permanently deformed to some extent. The point at which maximum strength is reached and the plastic begins to deform is the yield point. Maximum strength at yield will also often be quoted by suppliers to denote this.

Toughness: Toughness is a measure of how much *energy* a sample can absorb without breaking. (The strength was a measure of how much *force* a sample can absorb.) Put another way it is the ability of a material to cope with shock, such as impact, and deform without breaking. A simpler way of thinking of toughness is as a combined property of both strength and plasticity.

Elasticity: This is the ability of a material to return to its original shape after deformation. The point at which these properties are lost (e.g. when an elastic band breaks rather than stretching any further) is called the elastic limit.

Ductility: is a classification of a material showing plasticity (plastic behaviour). A ductile material can be plastically deformed without material failure beyond the yielding point. Ductility and plasticity are similar concepts but a material can show plasticity without necessarily being ductile. Polycarbonate is an example of a ductile material.

Fatigue strength: When a plastic component will be used for a long time under load it may be necessary to test performance over long periods of load exposure. There are two types of fatigue generally denoted creep fatigue (for a static load) and dynamic fatigue (for a load which varies on and off for a specified number of cycles).

Rigidity: The resistance of a material to bending.

Hardness: Hardness can be defined as a measure of the ability to resist permanent indentation. A harder material will resist penetration better than a softer one.

3.4.1 Tensile properties

There are a number of tensile testing standards such as ISO 527 (Plastics – Determination of tensile properties, in several parts).

Tensile strength properties allow manufacturers to compare the strength of their materials and this allows direct comparison of materials which have been tested to the same exacting methods. Tensile strength measures the maximum strain a material can withstand and therefore is a measure of the material's resistance to being pulled apart when placed under a tension load. The test method itself consists of placing test pieces of standard size and thickness (characterised by their dog-bone shape and therefore referred to as dog-bone samples) between the grips of a pulling machine. Specimens are shown in **Figure 3.2**.

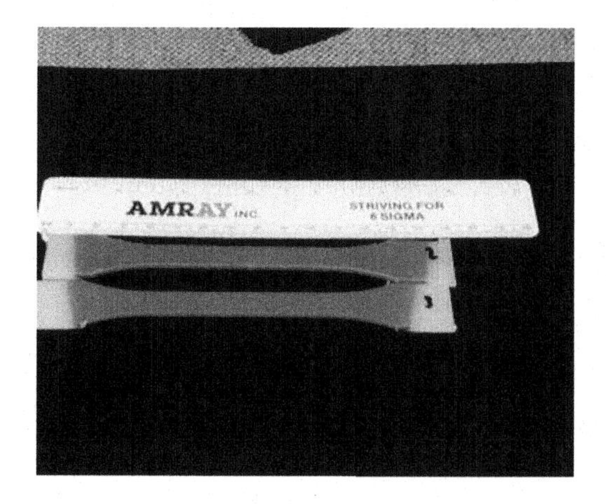

Figure 3.2

Dog-bone tensile specimens

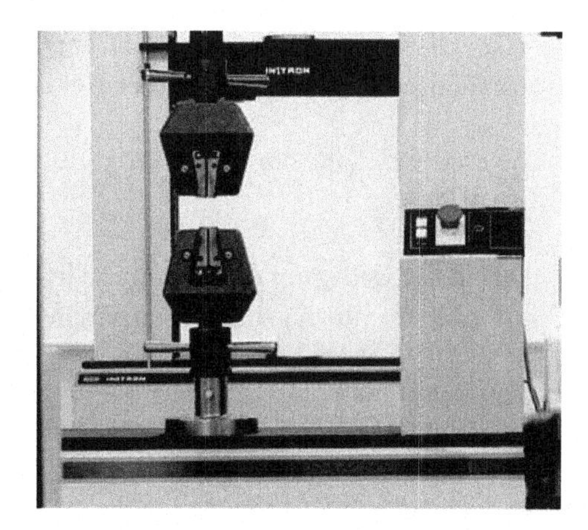

Figure 3.3

Tensile test grips in an Instron universal materials testing machine

The sample is clamped between two grips, as shown in **Figure 3.3**, and then pulled apart (stretched) at a specified rate (2 mm/min, 5 mm/min etc) by moving the 'jaws' of the machine away from each other. This rate of pulling is especially important for plastic materials due to their viscoelastic nature (see Part 1). For this reason samples pulled at different speeds cannot be compared. These machines tend to be computer controlled and calculate a variety of material property values based on the measurements of the amount of force exerted on the sample. By dividing the exerted force by the cross-sectional area the stress can be calculated. By continually calculating this force (stress) and strain relationship, graphs of the material's performance can be generated.

This is generally done until either the sample breaks (maximum tensile strength has been reached) or the specimen has reached the maximum strength but behaved plastically and yielded. The sample continues to be stretched until it breaks or a specific length of elongation relative to the original size has occurred. This happens with materials that are very ductile. The strength in these cases still relates to the maximum stress that occurred during testing, even though the sample yielded and did not break at that point. The amount a material stretches relative to its original size is reported as the elongation of the material. The elongation is dimensionless.

These kinds of stress-strain curves can be generated for any material, not just plastics. The shape of the curve will be dependent upon the properties of the material being tested, and also allows other properties to be calculated as well as just the tensile strength.

As the material has undergone stress and stretched a stress-strain curve can be generated. As the amount of stress has been increased, there has been a corresponding elongation of the test specimen. So the stress-strain curve represents the force from the machine divided by the cross-sectional area of the specimen on one axis, and the resulting elongation of the specimen relative to the original sample on the other axis. The gradient of the curve of the stress-strain plot is known as the tensile modulus and has the same units as the strength. A steep gradient means a high modulus (resistance to deformation), a shallow gradient means it deforms easily under load.

From the area under the curve, the toughness of the material can be calculated (a combination of strength and plasticity).

By using standardised samples, speeds, and loads, the properties of various plastics can be compared with each other as shown in sample curves in **Figure 3.4**.

From **Figure 3.4**, material A exhibits the highest stress (therefore has the highest tensile strength). It has broken at a low strain rate (low elongation) and is therefore glassy and brittle. The gradient is steep, therefore it is a high modulus material, but is not very tough as the area beneath the curve is not large.

Material B is relatively stress resistant, but is not as strong, nor did it break like material A. At the point of highest stress the material yielded (plastically deformed), it then continued to yield until it fractured. The gradient is shallower than material A, so the modulus is lower but this material is much tougher than material A as the area beneath the curve is greater. It has better elongation than A but not as good as C (maximum strain).

Material C resists far less stress than either A or B, so is the weakest of the materials here, but behaves plastically and yields, it then continues to yield until it reaches maximum strain and the test is stopped. The area beneath the curve is large, so this material has some toughness. The gradient is very shallow so the modulus of this material is very low.

A comparison of strength data for four materials is provided in **Table 3.4**.

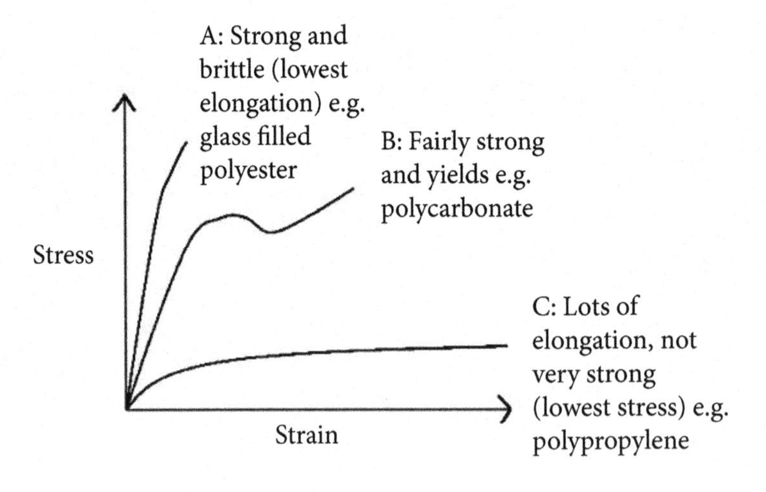

Figure 3.4

Typical stress-strain curves of different plastics

Table 3.4
Tensile data of selected plastics (ASTM D638 - 08)

	PP	PS	PMMA	Epoxy (reinforced with glass)
Tensile strength *(N/mm^2)	21	45	70	30 (965)
Tensile modulus *(N/mm^2)	1,100	3,200	2,700	21,500 (39,300)
Elongation %	>100	3	5	4
* *There are several units which may be used depending on the test methods. For tensile strength a Mega Pascal (MPa) may also be used. 1 MPa = 1 N/mm². For modulus the Giga Pascal (GPa) is often seen and is equivalent to 1000 MPa*				

Polypropylene (PP) is a semi-crystalline material. Its behaviour is similar to material C in **Figure 3.4**. It has relatively low strength compared to the other three materials, is the least rigid (has a lower modulus) but is very ductile (can be elongated). These properties explain why PP is used mainly in low cost items with low technical requirements such as packaging. (It is also very cheap and easy to process, which also helps!). Another very ductile material is nylon 6, which can be elongated well in excess of 100% its original length. (This is not elastic behaviour, as the deformation is permanent. With elastic behaviour the sample would return to its original size after the test). Ductility is a useful property especially for assembly operations and allows materials such as PP and nylon to be used for push fit applications.

A brittle material like PS will simply shatter when pressure is applied. This is an amorphous material (and it is therefore transparent) and brittle. This brittleness is illustrated by comparing the elongation values with polypropylene. Whereas PP can stretch by more than 100% of its original length without breaking, PS fails (the material breaks into two) at just 3%. Poly(ethylene terephthalate) reinforced with 30% glass fibre has similar elongation properties to PS, in this case the glass fibre limits the amount the material can stretch before breaking, but boosts the load-bearing capabilities of PET alone.

Another amorphous material is PMMA and again this is a transparent material, more commonly known as Perspex. Compared with PS this material is much stronger (tensile strength) and ductile (elongation) but slightly less rigid (modulus). It should therefore not be too surprising that this material is considered an engineering plastic as opposed to a commodity plastic such as PP and PS.

The final of the four materials here is a thermoset epoxy resin. These materials are characterised by their chemical crosslinking and this gives rise to a much higher rigidity (modulus) than any of the three thermoplastic materials. It is still a brittle material similar in elongation properties to the amorphous materials. The strength properties of this material on its own may not look that impressive, however if glass or carbon fibre is incorporated into the manufacturing process, a component of considerable mechanical strength and rigidity can be produced. This is a composite. The enhanced properties are shown in brackets.

3.4.2 Flexural testing

By changing the direction of the load and the jigs within a universal testing machine, a number of other tests (and stress-strain data) can also be generated along similar lines. For example by bending the sample, rather than stretching it, the flexural properties can be calculated. By compressing it (the opposite direction to a tensile test), the compression properties can be found. In all three cases (tensile, flexural and compression) the strength is that needed to break the sample (or the maximum stress reached if it does not break).

Flexural tests are very useful for designers interested in the stiffness of materials. They are often carried out on the same machines that generate tensile data, however instead of gripping each end of a sample and stretching it as in tensile tests, this method employs a three-point or four-point bend. A three-point bend can be simply envisaged by holding a ruler at both ends between two fingers and asking someone to the press down in the centre of the supports (you) causing the ruler to bend in the centre. A jig for testing this is shown in **Figure 3.5**. A four-point bend involves two specified points of pressure on the supports. Materials with poor flexural strength are easily snapped in half. Lead pencils are a common example of a product with weak flexural properties.

Since a three-point flexural test is like testing how much weight you can put on a bridge, it is perhaps not surprising that this test is very useful for looking at the structural properties and applications of materials. The amount of flexing (deflection) required by the sample is limited and this is not an indefinite flexing test. Therefore, many flexible materials simply do not break during this test.

Like tensile testing there are specified methods and specimen sizes (ASTM D790, ISO 178) and data usually generated from these tests include flexural strength and flexural modulus. The units are as with tensile properties. Two materials with very differing flexural properties are PP and 30% glass fibre filled PET. PP, as used for the living hinge mentioned in Part 2, needs to flex over and over again without breaking. When testing PP samples by flexural testing they would simply deflect under the weight on them but not break. The flexural strength would be low (as they flex easily). In comparison, the glass filled PET would have a high flexural strength (it would resist flexing under load) but at a certain point of deflection would break.

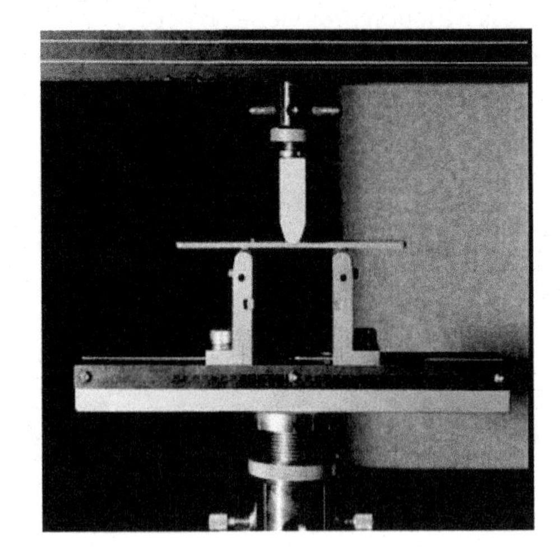

Figure 3.5

Three point flexural testing

To return to PP and the living hinge application, typically used for caps on food and toiletry packaging, this is designed to flex many times during service without breaking. However, in this example, it is not too catastrophic if the hinge eventually breaks. In other cases of repeated flexing, such as a plastic bridge, the consequences for failure could be more severe. This is an example of a case where fatigue testing could be utilised to test the material through a high number of flex cycles to see how it will perform in the long term. Fatigue testing can relate to any tests and test environments where the material undergoes fluctuating loading.

3.4.3 Compression testing

If a force is applied in the opposite direction to a tensile test, the specimen is compressed. Like tensile and flexural tests, stress-strain curves can again be generated to predict behaviour. Compressive strength is given in MPa. Compression testing (e.g. ISO 604) generally has two useful applications. For elastomers it is useful for determining the point at which elastic properties (elastic behaviour) are lost and the material deforms permanently (plastic behaviour) . This is called the compression set.

For rigid plastics and composites (brittle materials), this test can provide useful mechanical information on the maximum tolerance in compression at low rates of stress and strain before material shattering occurs. (The test is of less significance for ductile materials.)

Plastics are rarely used in applications where they will experience significant compressive loads, as these are normally reserved for structural materials like metal and concrete, and therefore this property is not of great design significance. However it is wise to avoid the likelihood of shattering a brittle material, potentially causing massive material damage to a component or assembly. Compression testing can provide information on compressive strength, compressive yield strength and elastic modulus, and with other mechanical data this can be used to assess overall mechanical strength.

A plastic block under compression testing is shown in **Figure 3.6**. Compression tests can be done on the same machines as tensile and flexural testing but with compression jig plates fitted. The direction of the movement of the plates is the same as with flexural testing. The plates are bought together compressing the material in the middle.

Figure 3.6

Compression testing

3.4.4 Impact properties

Impact properties measure how a material may behave if dropped or hit. Examples of this are a plastics drinks bottle dropped from a table and impacting with the floor, a plastic automotive bumper in a high-speed impact collision or a motorcycle helmet hitting a road. The impact test is a measure of the energy absorbing properties of a material and an example of vehicle impact testing is shown in **Figure 3.7**.

As mentioned previously, in tensile testing the equivalent ability to absorb energy before breaking (toughness) can be measured by the area under a stress-strain curve as shown in **Figure 3.8**.This was generated from a force divided by the cross sectional area (stress) and the deformation (strain). It is therefore not surprising that the units of impact relate similarly to impact energy on a known cross sectional area and a deformation. Common units are kJ/m^2. However unlike tensile tests, impact testing measures a one-off controlled impact of a known energy. There are a number of internationally agreed standard methods and specimen sizes (ISO 179, ISO 180, ASTM D256, ASTM D3763, and ISO 6603). The results indicate how resistant a material is to impact damage. Perhaps the best known impact tests are those crash test dummy impacts used to model and measure the behaviour of humans in automobile crashes.

The impact properties of plastics can vary greatly, from materials with little impact resistance to materials such as Kevlar that can resist ballistic impacts and are used in bulletproof vests. Fillers and additives can also be used to greatly modify the impact properties.

Figure 3.7

Automotive front bumper under impact test. The rig is set up to monitor impact and the bumper will be hit at the centre at a known test speed

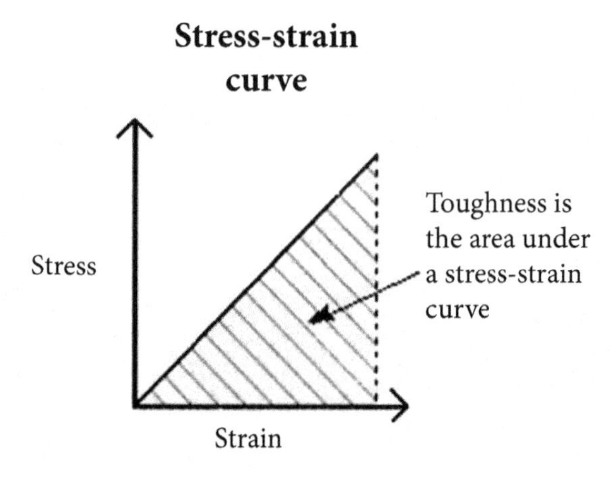

Figure 3.8

Toughness indication from a stress-strain curve

There are common plastic materials well known for giving good impact protection, and these are used in numerous such applications where these properties are required. The expanded polystyrene mouldings that package and cushion the contents of boxes during transport and delivery are a commonly seen example. Interestingly, in its unexpanded form, polystyrene is very brittle and easily damaged by impact. The reason expanded polystyrene is effective is because of the cushioning effect of the trapped air within the plastic. These cells collapse individually, like a stack of mini balloons absorbing the impact. This effect can be seen on a larger 'balloon' scale in polyethylene film used in Bubble Wrap®. This is actually a brand coined by the inventors and patented by the company Sealed Air, but the term is commonly employed for all similar products (see **Figure 3.9**). The air acts as an impact resistor and cushions any damage. The air bubble needs to compress but not pop for maximum efficiency, (however subsequently popping this material is a popular pastime!)

From the behaviour of a material during impact failure information can be gained about how brittle or ductile it is. There are four possible outcomes for materials experiencing an impact:

Yielding: similar behaviour as seen in materials yielding under tension, this leaves permanent deformation. The impact sample will show no cracking behaviour.

Cracking: the material shows a pattern of cracks (without permanent yielding) but retains its shape.

Brittle behaviour: The material shatters without yielding. Polystyrene is a good example of this behaviour. These are very glass-like materials.

Ductile behaviour: Cracking and yielding together. Polycarbonate is a good example of a material that behaves in this way and is used in motorcycle helmets to reduce injury during impacts. The good impact strength relative to other common plastics can be seen in **Table 3.5**. This behaviour is also the reason that once damaged, helmets no longer offer continued levels of protection. Once a material has yielded, the ultimate strength on subsequent impact is substantially reduced.

Figure 3.9

Packaging material using trapped air for impact cushioning

Table 3.5
Izod impact strengths for common materials (ASTM D256 - 06ae1)

Material	Izod impact strength (J/m)
Polypropylene	80
Polycarbonate	800
Polystyrene	28
Poly(ethylene terephthalate)	70

3.4.4.1 Methods

Generally impact tests are based on one of two principles, either a pendulum striker is used on the plastic sample or a weight is dropped on a sample.

In the pendulum tests, the user releases the load, which swings and strikes the plastic sample centrally between sample supports. The amount of energy required to break the test piece is recorded.

These are Charpy and Izod type impact tests. These tests specify striking energies and sample sizes and if notched, the notch size is specified. With naturally non-brittle materials, notching samples with a 'v' shape (cut in the specified manner to a standardised size) encourages the sample to break by raising stress concentrations.

Only when samples break can useful data be collected. Notching the specimens allows both an increase in the accuracy of data, and is also the site of impact damage propagation (the notch face) to be known. This is because notching encourages brittle failure of materials as it raises the stress of the material at the area of the notch.

There are differences in the clamping of the material, the specimen size and the striking energies used in Charpy and Izod test methods, and therefore samples tested by different methods should not be compared.

3.4.4.2 Charpy impact

In this test method (ISO 179), the sample is placed horizontally between two retention blocks and the pendulum is released to strike the centre of the sample. Older machines contain an indicator scale above the specimen for the operator to directly read off the impact energy used. In newer machines, this value is given on a digital display for the user. In the case of ISO 179 the question whether the machine is instrumented, or energy manually read off a scale by the user determines which part of this standard is used. This test gives the impact energy at break. Consumed energy relative to the pendulum should be between 10 and 80%. In ISO standards the Charpy impact strength is expressed in kilojoules per square metre (kJ/m^2), this is the impact energy relative to the central cross sectional area. Specimen behaviour can also be reported using one of the following five descriptions: no break, partial break, tough, brittle or splintering. A range of pendulum striking energies are commonly available which give between 0.5 and 5 Joules of impact energy, although pendulums of up to 50 Joules are used for very impact resistant materials.

3.4.4.3 Izod impact

This method is like a Charpy test in that it measures the energy absorbed by the sample from a single impact with a pendulum of known weight and velocity. In this test method (e.g. ISO 180) the direction of material clamping is vertical, and a smaller length sample is used. An Izod impact test is shown in **Figure 3.10**. Again the units are kJ/m^2 and relate to the energy absorbed relative to the original energy generated by the swing of the pendulum. The pendulum is generally used at a standard striking energy of between 1 to 5.5 Joules dependent on the strength of the sample. The absorbed energy is displayed (instrumented machine) or can be read off the scale (manual machine).

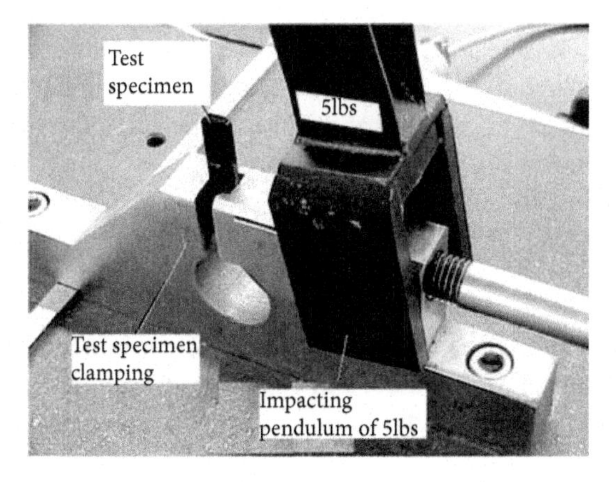

Figure 3.10

Izod impact testing

Like many of the other tests described within this section, other standards specify different units for reporting Izod impact strength. For example ASTM impact energy is expressed in J/m or ft-lb/in.

The Izod impact strengths of several common polymers were given in **Table 3.5**.

3.4.4.4 Falling weight impact testing

Charpy and Izod tests are examples of pendulum impact tests, another way of measuring impact is to drop free-falling weights onto a sample. Greater weights and energies can be applied with these methods if data on larger impacts are required. In this type of test the energy required for failure, rather than the absorbed energy is usually presented.

One example is the falling dart impact test (ASTM D5420). In this test a flat, rigid specimen is impacted by a striker from a falling weight. The energy of impact is calculated by a simple multiplication of the weight and height of the dropped striker, and is expressed in Joules. Samples are usually placed below the striker on a flat surface and upon impact the striking energy is transferred to the sample. Results depend on a number of parameters including sample geometry, the geometry of the falling weight and the height it is dropped from. As such these test types serve as comparators rather than providing the absolute values that can be obtained from Charpy and Izod impact tests. Certainly any deviation in geometry invalidates comparison from sample to sample. The major differences between pendulum and falling weight impact tests lie in the fact that samples in the latter are flat rather than a supported beam, and also that there is no notch or stress raiser in these tests.

3.4.4.5 Instrumented impact testing

This method (ISO 6603) has much in common with falling weight tests but more data can be obtained using these machines. It is necessary to ensure sample failure with each single impact strike, and a load-deflection curve is generated that allows the mode of failure to be determined (using both observation and data examination). A variety of data can be extrapolated from the load-deflection curves including toughness, total energy, energy at a maximum force and energy at break. Like tensile stress-strain curves, it is possible to see modes of failure from such curves. In these tests a plunger can be dropped at a high speed of 200 m/min. The energy of impact can be varied, by varying either weights or impact speeds to ensure that on hitting the surface sufficient velocity is retained to penetrate the sample. The force (load) versus the sample penetration (displacement) is calculated and displayed electronically. Like falling weight tests, only results using the same testing parameters and samples can be compared.

With all impact tests, temperature can affect the material performance and therefore it may be necessary to test materials at a temperature that reflects the end use environment. This is especially relevant to thermoplastics whose impact performance can change significantly with temperature. Testing at service conditions gives a more reliable idea of materials performance.

A summary of the four major mechanical testing modes discussed so far is given in **Table 3.6**.

Table 3.6
Comparison of mechanical test methods

Testing Type	Overview	Direction of movement	What it tells you
Tensile	Sample pulled apart		How easily a material may tear or permanently stretch
Flexural	Sample deflected in central region		How flexible a material is – will it bend?
Compression	Sample crushed as plates move together		How will a material behave when squashed?
Impact	Sample held still and struck with moving weight		How will a material behave when hit or dropped?

3.4.5 Creep

During service, a component under stress may fail at actual stresses well below the measured mechanical properties. This can be attributed to fatigue, either by prolonged exposure to consistent load, often referred to as creep fatigue, or under a cyclic changeable load, often referred to as dynamic fatigue.

The reason for failure is that under long term loading plastics may begin to change shape and deform due to their viscoelastic properties. The phenomenon of a material deforming over a period of time due to an applied load is called creep. With this in mind, a solid state creep test (ISO 899) gives data on performance over a longer stress exposure, and measures the dimensional changes which result. Creep is dependent on the load, the time under load and the temperature, and these can be accurately measured by suitable tests.

However it should be noted that for many applications, while creep may be occurring, it is not really a design factor. It is only relevant for those components which are subject to loads for an extended time, either at room or elevated temperatures. So for a plastic drinks bottle, a mobile phone cover or a disposable plastic bag, creep does not really matter. However, if the plastic bag was being used continuously to carry things (subject to continuous loading) then creep would become a factor in its design. For a simple plastic bag you would see creep occurring as a stretching of the plastic handles over time (since they would be the weakest point).

For an engineering application such as a bearing, impeller blade or pulley there may be dimensional changes that affect service, and even lead to catastrophic failure of the part if it has not been correctly designed and specified. Since creep properties are temperature dependent, high temperature environments (for example under-bonnet) can induce higher levels of creep in materials.

For ISO 899-1, which looks at tensile creep, the tensile creep modulus is reported at 1 hour and 1000 hours at a strain less than 0.5% in the units of MPa. It is measured at various intervals between these times as specified and graphs of strain over time or stress over strain (as in **Figure 3.11**) can be plotted.

A further important application for polyethylene is in underground gas pipes. Here the pipes are constantly under load from the gas for decades of service life, however in this instance the load is internal. In this case it is important to know how well they will perform, and for how long they will perform under long-term internal pressure. ASTM D1598 - 02 Standard Test Method for Time-to-Failure of Plastic Pipe Under Constant Internal Pressure is such a test, and an example of how tests must be designed and selected for the intended purpose.

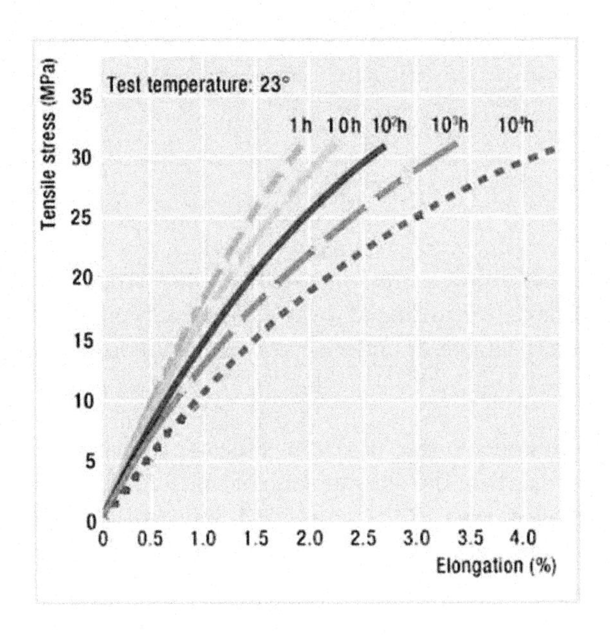

Figure 3.11

Example stress-strain curves measured in accordance with ISO 899-1, showing the effect of creep over time

An example of a dynamic fatigue scenario is a tensile test (or a component in service) where a load is applied every five seconds held for five seconds and then released. Dynamic fatigue is not as well studied, but many of the factors such as temperature, mean load and time of load are important in assessing material behaviour. The material response is more complex due to the relaxation period between applied loads (again viscoelasticity must be taken into account) and further discussion is beyond the scope of this book.

3.4.6 Hardness

To measure the hardness of the surface of a material a gauge called a durometer can be used. Various types of durometer exist for measuring hardness; a commonly reported measure is the Shore hardness scale measured with a Shore durometer.

Shore hardness durometers (ISO 868:2003) of relevance to plastics are those of type A (used to measure soft materials) and of type D (used to measure harder materials). The durometer is forced into the sample under specified conditions and the penetration depth is measured. For full penetration a hardness of 0 is reported, and for no penetration a hardness of 100. Therefore the indentation is inversely related to hardness. The time of penetration and the scale that was used (A or D) must also be reported.

A/15:55 represents a hardness using scale A, where the indentation was carried out for 15 seconds and gave a reading of 55. This could be representative of fairly soft materials such as polyurethane used for cable covers. D/15:92 on the other hand represents a much harder material, such as an epoxy resin used for making electrical appliances.

A number of other hardness tests also exist and data cannot generally be compared from one to another.

3.5 Thermal properties

There are a variety of tests that can be used to study heating effects.

Plastics can be exposed to thermal degradation both in processing and in service.

Processing stability can be measured by slowly elevating the processing temperatures. The temperature at which the physical appearance changes such as by yellowing or browning is indicative of the material's heat stability. However, as well as the temperature, the time of exposure must also be considered. A material may be fine at 250°C for 1 minute but may degrade after 5 minutes. The time a material is kept at a temperature is called the 'dwell time' or 'residence time', and some materials, such as PVC are particularly sensitive to long dwell times. For this reason PVC tends to be processed on special machinery that minimises the material's exposure to temperature. There are good reasons for this. As PVC degrades it gives off hydrochloric acid and can actually begin to corrode the very machines it is being processed upon. PVC can also be problematic in recycling for this reason (see Part 4) if other materials become contaminated with PVC waste.

Heat can cause a variety of changes in a plastic component. It could shrink, expand, bend (deflect), harden or soften depending on how it is used and how it is held in place. It is important a designer knows how a material will behave if it is not to fail in service.

The key material properties that need to be considered in these cases are the melting point, the crystallisation level, and/or the glass transition temperature. The thermal conductivity of the material is also a factor. The likelihood of component movement, whether by expansion or contraction as temperature changes also needs consideration.

Expansion and contraction can cause problems in service, especially if a plastic component is being used in conjunction with other materials. Consider the example of a plastic automotive body panel being used in conjunction with metal panels and fittings. The degrees of contraction and expansion of the panels will vary even with the relatively small variations of temperature found in the UK climate for example. Assemblies will often be designed to incorporate gaps, which will prevent any buckling occurring as parts expand against each other (these are not obvious unless you actually look for them – see **Figure 3.12**). As an example the coefficient of linear thermal expansion of aluminum is 23.6°C^{-1} x 10^{-6} compared to 145°C^{-1} x 10^{-6} for polypropylene.

Not surprisingly there are a range of standard tests used to study this behaviour, such as measuring the coefficient of thermal expansion and the heat distortion temperature.

3.5.1 Analytical thermal methods

There a number of analytical methods that can be used to determine when polymers go through transitions in state such as glass transition temperatures, crystalline melting and degradation. These methods require only very small (milligram) quantities of polymer or plastic to provide results. Differential scanning calorimetry (DSC) testing (ISO 11357) is commonly used to report properties such as the glass transition temperature. This method can provide very valuable information about a material by measuring energy changes. As polymers pass through various states heat is either absorbed or emitted. By comparing the temperature relative to a reference sample, it is possible to generate transition data across a programmed temperature range. This technique can also be used to monitor the onset of curing in thermoset materials. The various parts of ISO 11357 include sections

Figure 3.12

An automotive assembly showing expansion gaps between different materials on measuring melting points, glass transitions, crystallization and specific heat capacity using this one technique.

Thermomechanical analysis (TMA) can be used to calculate thermal expansion and glass transition temperatures (ISO 11359-2). Here dimensional changes are studied relative to temperature. The specimen is placed on a support and a probe placed on the sample which measures changes in length. The results are reported as a length change per temperature change.

Thermogravimetric analysis (TGA) is used to determine when a material begins to decompose on heating (ISO 11358). As temperature increases components of a material become gases and are released from the sample, resulting in a measurable decrease in weight. The degradation of polymers and plastics can be accurately measured using this technique.

3.5.2 Heat deflection temperature (HDT)

This is a measure of a material's performance under high temperatures and loads. An example standard is ISO 75. It is similar to a three point flexural test in that it measures sample deflection, but here a fixed amount of deflection is specified and the test is performed in an oil bath, which is gradually heated. The temperature is monitored by a thermometer, and a fixed force is applied to deflect the material. After a certain level of heat, the materials will deflect the specified amount and the user can read off the temperature at which this deflection occurred. Deflection is measured using a dial gauge or similar device and the temperature of distortion is read off a thermometer. The results are usually presented at two different loads (1.8 MPa and 0.45 MPa) and the data is the temperature of deflection in centigrade.

3.5.3 Vicat softening temperature (VST)

This test is similar to HDT and again performed in hot oil. Some equipment incorporates both tests on one machine. In this test (ISO 306), instead of applying a load across the sample, a standard indenting tip is used to measure penetration into it. Once the tip has gone in to a depth of 1 mm then the Vicat softening temperature has been reached and the corresponding temperature of the oil is read off the thermometer in °C. Two distinct loads are used (10N and 50N) and two distinct heating rates (50°C/h or 120°C/h) and therefore the test regime must be quoted.

3.5.4 Thermal expansion (coefficient of linear thermal expansion - CLTE)

Change in length or volume as a function of temperature can be measured by the coefficient of thermal expansion. The significance of contraction or expansion for an assembly has been mentioned previously. However, such changes can also cause problems for an individual component depending on how it is held in place or fixed. Extra stresses at these points within components (called internal stresses) can lead to material failure in extreme cases.

CLTE is defined as the fractional change in length or volume for a unit change in temperature. Compared with steel, unfilled plastics expand between two and ten times as much, however this can be reduced by the addition of fillers or reinforcements. The expansion is in the range 5 to 40 x 10^{-5} per degree centigrade for unfilled polymers, with the lowest values for polymeric materials being in thermosets (crosslinked materials). This may not seem a massive amount until you consider how much temperature variation there may be (even just considering seasonal variations in room temperature) and the very precise tolerances between components which are required for some specialised plastic applications.

CLTE can be measured by using accurate length measuring techniques (such as TMA) and good temperature control of specimens.

3.5.5 Thermal conductivity

Thermal conductivity gives a measure of how well a material can transport heat. Plastics have very low values, typically in the range 0.14-0.9 W/mK (see **Table 3.7**), and this has implications for plastic processing, which are described in Part 2.

Table 3.7
Thermal conductivity of various materials

Material (lowest first)	Thermal conductivity (W/mK)
Air	0.03
Expanded polystyrene (foamed)	0.03
Wood	0.04-0.4
Plastics	0.14-0.9
Rigid poly(vinyl chloride)	0.14
Water	0.6
Epoxy resin	0.88
Steel	17-50
Aluminium	211
Copper	370-390

In general terms, plastics are very poor heat conductors compared to metals, and they can therefore be utilised as insulators. However, as air is an even better insulator, the insulating properties of a plastic can be improved by having a large number of air gaps within it. This effect can be produced in plastics by foaming (described in Part 2). Expanded polystyrene (also referred to as Styrofoam) is a good example of this.

Even the better heat conductors such as epoxy resins are still used for their insulating properties. This material is used in the electronics industry and protects electrical components from short-circuiting. Epoxies are used in printed circuit boards and transistors as well as for components such as motors, generators, transformers and bushings. The conductivity of plastics can be improved by mixing in metallic fillers.

Thermal conductivity and related properties such as thermal diffusivity are covered under ISO 22007-1:2009.

3.5.6 Flammability

The behaviour of plastics if they reach very high temperatures and potentially ignite or burn in service must also be considered. Whether it is a house fire, an automotive fire, a factory fire, an electrical fire, or a fire in an underground pipe for instance, the behaviour of various plastic components will be of major importance to human health and safety.

At one extreme a material may simply smoke and self-extinguish when ignited, whereas at the other extreme another plastic may burn fiercely, with burning droplets that fall onto the surrounding area. One may be suitable for sofa cushions, the other not! The standard most often seen for plastics is the UL 94 test (Underwriters Laboratory). This test is applicable to plastics in devices and appliances, and tests small size samples. Therefore it would not be applicable to a sofa cushion, but would be suitable for a mains cover plate, or plastic components in a kettle. The test pieces are prepared using various thicknesses of plastics, in a series of standard colour versions. They are ignited by a Bunsen burner under controlled conditions and the flammability assessed. The rating of H (horizontal) or V (vertical) represents the orientation of the specimen under burning. V-0 represents the best flame retardancy with V-1 and V-2 getting progressively worse.

3.6 Electrical properties

3.6.1 Electrical conductivity

Like thermal conductivity, generally plastics are poor carriers of electric current unless doped with a suitable conductor such as a metal powder. This property is temperature dependent and hotter plastics become less resistant to electrical conduction. Plastics are used as insulating materials for many electrical devices and wires are commonly coated in plastic materials.

However, there are also speciality polymers that can conduct electric current. These materials possess many of the properties of metals (such as similar magnetic or optical properties) and are collectively known as conducting polymers. One such material is polyaniline. Materials such as these have become very important for the development of microelectronic products such as flexible displays, and may have conductivity values around 1.5×10^7 $(\Omega\ m)^{-1}$, which is in the same order as metals like silver at 6.8×10^7 $(\Omega\ m)^{-1}$, aluminium 3.8×10^7 $(\Omega\ m)^{-1}$ and stainless steel 0.2×10^7 $(\Omega\ m)^{-1}$. For comparison, an insulating plastic such as polyethylene will have conductivity around 10^{-15} $(\Omega\ m)^{-1}$.

3.6.2 Commonly seen electrical tests methods

With the exception of conducting polymers, it is for their resistance to the passage of electrical current that plastics are generally used in electrical products. Properties such as surface resistivity and volume resistivity are used to measure this resistance (BS 6233/IEC 60093). Arc resistance or tracking index (BS EN 60112) gives an indication of the ability to resist a high voltage and low current exposure close to the surface, and measures the time and voltage required for the material at the surface to break down and form a conductive path. Dielectric strength (BS EN 60243) measures how much electrical strength a material has to resist electrical voltage, expressed in Volts per unit thickness (in many ways this can be considered the electrical equivalent of a tensile strength). It should be noted that the term dielectric strength relates specifically to an insulating material, however in theory both electric strength and dielectric strength are measuring the breakdown of the material.

3.7 Other properties

A typical plastic data sheet will contain information on most of the properties discussed so far. However, as the scope of plastic usage is so large and there are literally thousands of standards relating to the very specific tailored properties of plastics, it is impossible to cover all tests for all applications. Therefore while most of the main properties that will be encountered are here, there are many specialised testing standards beyond the scope of this book. Referring to the designated standard is therefore recommended in these cases. To conclude the testing section a further limited selection of properties is introduced. These relate to practical applications where aesthetics, design, environmental and acoustic properties may be important.

3.7.1 Transparency

Transparency is a very useful property as it allows plastics to be used in applications where glass is routinely used. For example windows, screens, light covers, viewing ports and protective shielding. Not all plastics are transparent. As light passes from the air into a polymer it will be scattered at the interface, and the transparency of semi-crystalline materials is affected by the amount of crystallinity. If a material is essentially transparent, like polycarbonate, there are useful test procedures that can be carried out for quality control.

One such example is the measurement of the light transmittance of plastic sheeting (ASTM D1746 – 09). Good light transmission is required for glazing applications, where glass is replaced. Polycarbonate (PC) has good light transmission, (transmitting about 90% of incident light) as well as being tough. It also has fairly good resistance to heat, and is consequently used in glazing, automotive lenses and large containers.

Transparency is an aesthetic quality (as well as having practical applications especially in relation to replacing glass). Other aesthetic type properties include tests for gloss, colour, haze (the cloudiness of a transparent material) and surface roughness. While many surface quality standards exist, automotive manufacturers, amongst others, often have their own internal standards relating to acceptable visual aesthetic properties. This is because appearance is so closely linked with perceived aspects of 'quality'.

3.7.2 Shrinkage

Shrinkage has great relevance for design engineering tolerances, especially where parts need to be fitted directly together or fixtures and fittings are incorporated (by machining for instance) after moulding.

During polymer processing operations, the nature of a material (whether thermoset or thermoplastic) the production method and the reinforcement can all affect the properties of shrinkage. Examples of standards for shrinkage measurements include ISO 2577 and ISO 294-4.

In injection moulding for instance crystalline materials shrink much more than amorphous ones, and this is generally true for all processes. However part thickness, the geometry of the mould tool or extrusion die, and the processing conditions will also all affect shrinkage levels to some degree. The dimensions of a mould tool are usually bigger than those required of the final component manufactured within it, and dimensional tolerances can be set with this in mind. Shrinkage tends to be different in the direction of polymer flow and the transverse direction. Reinforcement with fibres or fillers reduces shrinkage values overall, but also tends to produce even more anisotropic variations in components, with differing levels of shrinkage in different directions, and can also induce warping of a component.

An example of a material with a high shrinkage value is acetal (2%), which is crystalline. This shrinks roughly three times more than amorphous polycarbonate at roughly 0.6-0.8%. By adding 30% glass fibre, the shrinkage of the polycarbonate can be roughly halved again. The mould shrinkage is reported as the percentage change in dimensions relative to the original mould tool cavity dimensions.

3.7.3 Light stability and weathering

Exposure to sunlight causes degradation of plastics, due to the effects of both light and heat. For coloured plastics, exposure to light can cause a fading or changing of the colour pigments over time. This is commonly seen even in the UK as 'greying' of white garden furniture or the 'yellowing' of white PVC window frames. In an old Ford Escort car once owned by the author many years ago, the dashboard actually began to 'chalk' even though it was an interior component, due to the effect of light through the windscreen. (Chalking is the appearance of a white powder on the surface of the plastic, which can be rubbed off.) At the time I was working for a masterbatch supplier, formulating colours for desert use, so I saw the effects of selecting the right pigments for service life (or not) for myself. A lesson I have long remembered. Pigments are rated for light stability from 1-8 with 8 being the best. To prevent colour fading in applications exposed to sunlight, it is necessary to use very light stable pigments. Stabilising additives are often added to mitigate the effects of heat and light and minimise degradation.

Anticipating the environment a component will experience during service life is a vital part of materials' selection. It is no good specifying a material that will meet your requirements initially but fail after ten minutes, if the required service life is two years. Service conditions could mean exposure to an outdoor environment (wind, rain, and sunlight), a man-made one such as pipework carrying corrosive solvents, or a high temperature one such as experienced in under-bonnet applications. In all cases the resistance of the material to its application environment would need design attention.

Returning to the natural environment, the effect of light exposure on a polymer is to initiate a process called **photooxidation**. The main factor causing deterioration in plastic components exposed to outdoor service is thought to be ultraviolet (UV) radiation.

As shown previously, heat ageing and stability can be simulated using an oven and/or elevated processing conditions. Similarly, the effects of light exposure can be tested using either artificial light exposure in a laboratory (ISO 4892) or the samples can be exposed to the natural environment (ISO 4582, ISO 877).

If a product or sample is in a natural environment, exposed to the elements, the processes are referred to as weathering. Where that product or test sample is sited will affect its performance. There are standardised weathering sites situated around the world which include Florida and Arizona (both in the USA) and southern France. Each location has markedly different weather conditions, and all three are different to those experienced by the author in the UK. In looking at weathering data, the site that generated the test may be quoted e.g. Florida. Different climates such as hot and wet, hot and dry, temperate and cold all result in different weathering affects.

As previously hinted, the colour of plastic parts may also be significant. In black coloured polymers for instance the black pigment offers protection in two ways. It preferentially absorbs the radiation at the surface (over the polymer) but also blocks it getting deeper into the polymer matrix. This is why many plastics for outdoor use are coloured black to minimise the harmful effects of exposure to UV. Generally there is a blocking effect from adding stable pigments. There are also additive packages specially developed to protect the polymer from the effects of weathering exposure.

Whilst the combination of weather factors (moisture and temperature) undoubtedly contributes to material failure, they can each be isolated or combinations controlled better for evaluation under laboratory conditions. (If only the weather could also be so closely controlled.) A further advantage of laboratory weathering is that the degradation can be accelerated. To test the long-term exposure of a material in several sites in the world, samples must actually be kept on those sites for that length of time, and manufacturers must wait a long time for their results. Accelerated weathering in the controlled environment of a laboratory allows results to be generated much faster.

The effect of weathering on polymers can include chalking, crazing, reduction in mechanical properties and colour change. However impact strength appears strongly linked to the effects of sunshine hours alone. HDPE, commonly used as a crate material and therefore exposed to the elements will exhibit brittle cracks caused by UV light. This is known as environmental stress cracking and can be seen in other harsh environments (such as chemical corrosion) as well as related to weathering.

In all these tests any change in property can be compared with unweathered samples and therefore the changes in mechanical properties, dimensions and visual appearance can all be monitored over time.

3.7.4 Biodegradability

The various terms associated with degradation and biodegradability were introduced in section 1.9.1, namely non-degradable, readily degradable, controlled degradation and biodegradation.

A further consideration is that degradation caused by the weather, as discussed in the previous section, is different to biological attack by microorganisms such as algae, bacteria and fungi, and this leads to a further subset of degradation:

A **photodegradable** plastic is degraded by the action of natural light (exposure to daylight).

A **biodegradable** plastic is degraded by microorganisms in the environment.

A **compostable** plastic is one that degrades during composting. Inside a compost heap there is no natural light and compost heaps can get very hot. The environment is different to that experienced by a plastic biodegrading in a normal environment.

A further aspect of compostable degradation is that no toxic residues must remain after composting. Factors such as carbon dioxide liberation, oxygen demand and mechanical properties can all be used to measure degradation rates within a compost heap. Other methods exist but are beyond the scope of this book. The topic will be discussed further under the recycling of plastics in Part 4. Various standards exist to cover biodegradability, EN 13432: 2000 relates specifically to packaging.

3.7.5 Acoustic properties (effect of sound)

The acoustic properties of plastics are of interest to the design engineer for noise reduction applications. This can be achieved by two distinct mechanisms: sound reflection and sound damping, and both are necessary for noise reduction. As a sound wave hits a sound insulation barrier, part of the wave is reflected back and the rest is absorbed into the sound insulation barrier.

For sound to be absorbed it must be turned into heat by the absorbing material, and both elastomers and amorphous plastics have particularly good sound absorption properties. Metals on the other hand are bad sound absorbers. Foamed materials are poor reflectors of sound waves and they are therefore good for eliminating multiple reflections in soundproof rooms.

One way to measure acoustic properties is by measuring acoustic impedance. This is important in determining sound absorption, transmission and reflection and is defined as the material density multiplied by the velocity of the sound though the material.

3.8 Safety factors – a cautionary comment

The test data that is acquired using methods such as those described here cannot be thought of as absolute figures – perhaps in the way that data for materials such as metals can be approached. This is because the test data acquired for polymers depends on a whole number of factors, such as preparation of the test specimen, temperature, rate of loading, length of testing and others.

Whilst many of these issues are controlled to a greater or lesser degree depending on the test and standard method, even published data can vary depending on the test equipment, the manufacturer or the formulation. For this reason it is wise to consider all published data (and especially that used for marketing purposes) as approaching the upper limits of the materials' capabilities.

The reason for this is that a designer must be confident there is an adequate safety margin within test data to prevent premature failure. A simple example can be seen in the set of tensile testing strength data shown in **Table 3.8**.

Table 3.8
Example tensile test data showing variability in strength results between test specimens

Material A	Maximum strength (MPa)	Material B	Maximum strength (MPa)
Test 1	70	Test 1	100
Test 2	70	Test 2	40
Test 3	66	Test 3	110
Test 4	75	Test 4	35
Test 5	55	Test 5	70
Mean	**67.2**	**Mean**	**71.0**

On first glance material B has a better strength than material A. The mean tensile strength is 71.0 compared to 67.2. However, now imagine we were specifying a material to be used for a component where the maximum strength required was 60 MPa. In theory, both sets of materials exceed the specification. However, if we look more closely at the results for material A first, 1 of the 5 test results was below 60 MPa. Therefore if this was a genuine component we might expect that 1 in every 5 components could fail in service. Looking now at Material B, which looks stronger on average, in fact 2 out of every 5 results are below our specified maximum strength and could lead to a failure in service – imagine if that was a component on a bridge, a car or an airplane. Failure could be catastrophic.

Whilst in reality test data would not be expected to vary quite as widely as the examples presented here, there will always be a range of values. It is unlikely in a method such as a tensile test that all test specimens will perform exactly the same. There will generally be some deviation from the mean figure and this variability must be considered, as well as building in reasonable safety factors.

Part 4. The scope and applications of plastic materials

4.1 Typical applications of common plastics

This last part of the book will focus on sector-specific applications of plastics such as packaging, construction, medical, automotive, and electronic and electrical, considering a range of plastic materials. Please note that these sections are not all-encompassing and a material's omission from a particular section does not necessarily mean that it is not utilised in that sector. The aim is to cover a range of widely-used plastics and some typical applications in more detail. The section begins with a look at the packaging sector which accounts for 37% of all plastics consumed globally.

4.1.1 Packaging materials

Packaging materials are amongst the most common of all plastics and the bulk is made up of the following five materials:

- Polyethylene – (PE)
- Polypropylene – (PP)
- Polystyrene – (PS)
- Poly(vinyl chloride) – (PVC)
- Poly(ethylene terephthalate) – (PET)

For polyethylene and polypropylene the packaging market makes up the bulk of their use, whereas for PVC and PET, applications are spread into other market sectors.

Renewable materials (such as PLA – see 4.2.2) have niche packaging markets as do specialty plastics such as water-soluble PVAL. Since the vast majority of all packaging plastics are thermoplastic materials (with only minor use of thermoset materials, for example in the form of coatings), the processes most applicable are high volume, mass production, thermoplastic manufacturing processes. Therefore processes of interest to this sector include extrusion, injection moulding, blow moulding (both stretch and extrusion) and thermoforming. Packaging materials are frequently single-use, and generally required to be of low cost.

A further aspect of packaging is decoration. It is necessary to decorate a product to attract a customer to buy it. A range of decoration techniques are used by the packaging industry including labelling, in-mould product labels, printing, and self-colouring. Because of the high use of polyolefins, which are very difficult to stick to, surface treatments are necessary to prepare the surface for adhesion in some applications. Techniques such as corona discharge, gas plasma and flame treatment are used. A further option is to fluorinate the surface.

4.1.1.1 Polyethylene (PE)

Polyethylene as a material can be quite confusing as there are many different types. LDPE, HDPE and LLDPE have already been described in Part 1, however there are a number of other variations possible such as medium density polyethylene (MDPE) and ultra high molecular weight polyethylene (PE-UHMW). In 1939, when polyethylene was first available in the market, it was just called polyethylene or polythene for short. However as more and more control was achieved over branching and chain length with the introduction of the Ziegler-Natta process, this first material became known as LDPE, and other polyethylenes such as HDPE and MDPE followed on into the commercial market.

The next big breakthrough occurred in the 1970s when a new production technique was utilised, one which used metallocene catalysts. This led to the introduction of linear low density polyethylene (LLDPE).

The range of polyethylene materials makes it the largest volume polymer produced worldwide. Since it has a very simple carbon and hydrogen composition, its success could be attributed to this. However it is also very low cost and easy to process at relatively low temperatures. As well as this, our modern polymer production techniques can produce a whole range of tightly controlled products (see **Table 4.1**). It is possible to produce very soft rubbery polyethylenes such as VLDPE, right through the entire spectrum to ultra-strong ones (PE-UHMW).

As the density of the PE rises (see **Table 4.1**) so the tensile and flexural strength also increase, whilst the elongation decreases. The stiffness and hardness also increase. Similarly, as the molecular structure moves from multiple branches to linear chains there is an increase in tensile, flexural, stiffness and hardness properties. VLDPE and plastomers are very soft and rubbery as a result of multiple branches, whereas PE-UHMW is extremely strong and tough due to its linear and unbranched formation.

Table 4.1
Range of polyethylene densities available from various process routes

Production process	Density (kg/m^3)	Materials
High pressure	915-935	LDPE
Ziegler-Natta	900-965	LLDPE, HDPE, MDPE
Metallocene	880–940	Plastomers, LLDPE, PE-UHMW, VLDPE

In the film market PE is used extensively for flexible food packaging, agricultural film, bin liners and rubbish sacks. In the blow moulding market HDPE is used in plastic milk bottles, buckets, drums and bulk storage containers. The recycling of milk bottles will be discussed later in section 4.2.

4.1.1.2 Polypropylene (PP)

Of the various types of polypropylene available (atactic, isotactic, syndiotactic – see Part 1), it is isotactic material that is the most commercially important, although in fact the isotactic material used in most volume actually contains a mixture of both isotactic and atactic materials, giving it a 'broad tacticity distribution'. It has a higher degree of crystallinity than the other types of PP, which gives higher mechanical properties and better processability. Like polyethylene, metallocene catalysts can be used with polypropylenes to create different structures.

PP is widely used in injection moulding and blow moulding applications. Medical packaging, bottles, containers and caps are common products. PP can also be easily orientated and stretched, and film is biaxially orientated to improve a number of properties including gloss, clarity, tensile strength, water vapour barrier properties and resistance to oils and greases. This material is frequently referred to as BOPP. There are two methods for producing BOPP, ribbon extrusion and tubular extrusion. To produce the orientation it is necessary to draw the material in two directions (as opposed to just one in extrusion) both in the direction of flow and across it. Biaxially orientated film (BOPP) is used in over wrapping applications. This is the thin transparent plastic layer seen for example on a box of chocolates, a packet of cigarettes or a stack of CDs that seals the product.

As a material therefore it is versatile and low cost, and it can be foamed to give expanded PP. It can also be easily coloured using pigments. All these properties help to make it an ideal material for packaging applications.

Compared to polyethylene, PP is stiffer which makes it more suitable for caps and screw-type closures. It also has better heat resistance as the melting point is higher, and better transparency than LDPE. Its ability to flex repeatedly without failure also allows the production of living hinge type features as illustrated in Part 2.

Whilst PP is crystalline and rigid, blending with thermoset rubbers creates thermoplastic olefins (TPOs) and thermoplastic vulcanates (TPVs), and these materials can be used when a higher flexural modulus is required. PP can also be easily modified using a whole range of additives such as introduced in Part 1. This allows PP to be utilised in other, more technically demanding sectors such as the car industry.

The main drawback with polypropylene is the difficulty in adhering to its surface due to low surface energy (polyethylene has a similar low energy), which necessitates surface modification prior to many post decoration techniques. This can influence the choice of decorating method – pigmentation of the plastic (self-colouration) is one option.

4.1.1.3 Polystyrene (PS)

Polystyrene is a crystal clear and inexpensive amorphous resin often referred to a crystal PS. It is easily extruded, moulded or formed at relatively low temperatures. Polystyrenes are used to make a range of everyday goods including CD and DVD boxes, cutlery, yogurt pots and clear containers. However, as anyone who ever dropped a CD case will know – it is stiff but brittle. It also has poor barrier properties and limited solvent resistance.

When styrene is expanded using gas or carbon dioxide, goods containing only 5% PS can be produced (the rest being air). These very light foamed materials are used in the production of protective packaging, clam shell containers, cups, meat trays and egg cartons, and this is the most common packaging foam used. It has excellent insulation and cushioning properties in this form. This material, known as expanded polystyrene (EPS) will be discussed further in the construction section (4.1.5).

There are two other materials based on styrene that have overcome the problems of brittleness, and these are HIPS and ABS. These will be mentioned in the electrical and electronic sections (4.1.4).

4.1.1.4 Poly(vinyl chloride) (PVC)

PVC is both water and flame resistant and it is available in two versions: flexible and rigid. It is an inexpensive material and it is used in the construction and medical sectors as well as for packaging applications in film and mouldings. Packaging such as bottles and tubes for cosmetics and shampoos as well as films for packing diary and fresh produce can all be produced. More information on this material can be found in the construction section (4.1.5).

4.1.1.5 Poly(ethylene terephthalate) (PET)

This material is a polyester, and it was originally developed as an alternative to cotton, having improved crease and moisture resistance. Its specific use depends on whether it is produced as a high or low viscosity resin.

Low viscosity resins are used for bottles. They can also be used to produce photographic film, oven-ready food trays and packaging for toiletries and household products. In fibre form low viscosity resins are used for clothing.

Higher viscosity resins are used where higher performance is required such as in seat belts, tyre cords, and for the production of industrial fabrics.

PET is available in both amorphous (transparent) and semi-crystalline (opaque and white) versions.

Semi-crystalline materials possess high strength and stiffness, and dimensional stability, good slip and wear properties, electrical properties and chemical resistance, and a hard surface.

Amorphous PET is highly transparent, tough, and resistant to stress cracking, with low shrinkage and high dimensional stability. Amorphous PET is probably the more familiar to the consumer, as it is the material used for carbonated drink bottles. It has good chemical resistance and gas barrier properties to both carbon dioxide and oxygen that make it ideal for carbonated drinks packaging. It is also resistant to biochemical attack and environmentally benign.

4.1.1.6 Poly(vinyl alcohol) (PVAL) – a water soluble polymer

PVAL is also sometimes referred to as both PVOH and PVA, although strictly speaking PVA should only be used for polyvinyl acetate, see below. However this is also known as PVAc!

Many readers will have come across PVAL in their day-to-day life, perhaps without realising what the material was. The water-soluble tablets and capsules of detergents for washing machines and dishwashers are all packaged in wrappers made of PVAL (see **Figure 4.1**). They simply dissolve in the water, exposing the detergent material. Medicine tablets can also be coated in PVAL. Other lesser known uses include the production of golf balls that can be hit from ships and then harmlessly dissolve in the sea, and bags for dog owners to pick up dog waste which can be flushed away into the sewage system.

PVAL has very unusual properties. It is a synthetic polymer not a biopolymer, but water-soluble and therefore naturally biodegradable. It is also not produced from a monomer as most other plastics are but from another material, called poly(vinyl acetate) (PVA) by hydrolysis. The monomer vinyl alcohol does not exist. Most commercial PVAL actually consists of a copolymer of PVAL and poly(vinyl acetate) as the poly(vinyl acetate) is not all converted to PVAL. PVAL exhibits a number of useful properties which allow it to compete in niche markets including high tensile strength, good oxygen barrier properties, flexibility and good adhesion. However all these properties are dependent on humidity, as PVAL becomes more flexible in a moist atmosphere, and more brittle when dry. It is also used as an emulsifying[a] additive.

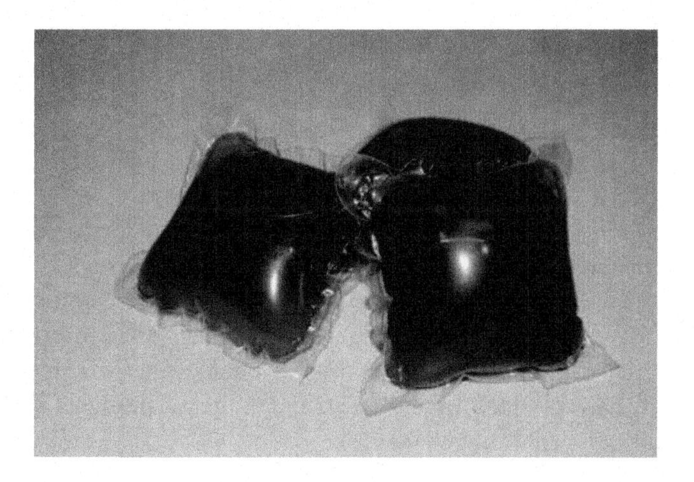

Figure 4.1

Machine-ready washing detergent capsules

a. ***Emulsifying***: *An emulsion is a mixture of two liquids that normally cannot mix (such as oil and water). When mixed together they would quickly separate if it were not for the action of an emulsifier. This serves to link the two liquids and stabilise an inherently unstable mixture. Examples of emulsions are paints (emulsion paint) and other coatings.*

Figure 4.2

Coextrusion blow moulded multi-layer sauce bottle

4.1.1.7 Ethylene-vinyl alcohol copolymer (EVAL) - oxygen barrier

This is a polymer with excellent oxygen barrier properties. It is often used in multi-layer blow moulding applications for this reason. EVAL is highly crystalline and therefore highly resistant to oils, solvents, odours and flavours. However it is also moisture-sensitive which is why it is used in internal layers in multi-layer blow moulding. Coextrusion blow moulding allows the production of bottles with different structures incorporating a barrier layer, for a range of hot and cold fill products such as fruit juice, ketchups and sauces (**Figure 4.2**), and chemicals.

Adhesive tie layers will also be necessary, which must satisfy the criteria for extrusion blow moulding, having suitable rheology to stretch with the other layers during processing, as well as forming a bond between the outer layers and the EVAL. (There may also be an internal layer of regrind incorporated in the structure.) If these bottles are recycled, the recyclate will contain proportions of all the various materials used in the original structure.

A further example of the use of multi-layer structures can be found in the automotive section.

4.1.2 Medical products and devices

Medical plastics include those used for equipment and medical devices such as syringes, tubing and blood bags as well as medical implants and diagnostic devices that are compatible with the human body. Specialist materials such as PVAL (see 4.1.4.) can also be used in items that will dissolve over time – for example in medical sutures. The use of plastics in the medical area continues to expand as they replace metals and glass products in disposable and multi-use products as well as finding new uses in implants and biocompatible materials.

A further aspect of medical polymers is the size of the devices. Often medical devices need to be very small. Developments in the use and understanding of nanomaterials as well as advances in micro-manufacturing processes such as micro moulding, allow the wide-scale manufacture of smaller and smaller medical devices. A large proportion of micro moulding is focused on the medical sector.

Bone is a natural composite material with mineral reinforcement in a collagen matrix. Similarly, plastic composites can be used as orthopaedic implants as bone and joint replacements. For example knee replacements can be manufactured from HDPE reinforced with carbon fibres. Polyethylene is used in medical prostheses such as hip, knee and shoulder implants. Other important grades of medical plastics include PE-UHMW (as joint materials), PP (for finger prostheses), PTFE (to knot arterial grafts) and acrylics (for contact lenses and bone cement).

PVC, in its flexible form, is widely used as a medical film for filtration membranes and in neurosurgical implants. It can also be used to store blood (blood and blood product bags) and for catheters. As a commodity plastic PVC is cheap, which makes it difficult to replace on a cost basis. It is also used for packaging medicines.

4.1.2.1 Polyurethane (PU)

Polyurethane is available as a thermoset, a thermoplastic and a thermoplastic elastomer material, and it is one of the most bio- and blood-compatible materials known today. Thermoset forms are used to coat implants. Linear thermoplastic polyurethane can be considered as similar in performance to PA 66, however this PU can be tailored to be rigid, flexible or tough depending on the formulation used. The thermoplastic semi-crystalline form of the material is not of particular interest in the medical sector.

Thermoplastic polyurethane elastomer (TPU) was the first thermoplastic elastomer to go into production. These materials are highly flexible, with high strength and elongation properties. They are also highly resistant to solvents, oils and many fats. As they are non-toxic and compatible with blood they are highly useful materials in medical applications, where they can be extruded, injection moulded or blow moulded. TPUs can also be used as media for tissue growth in implants, in the form of porous materials which offer anchorage for new growth. Further uses include components in artificial hearts, dialysis membranes and peristaltic pumps. Polyurethane is also commonly used for medical tubing. It is abrasion resistant, transparent and flexible, and can be sterilised.

The properties and processability described above also mean that TPUs can be used for wire and cable sheathing, and in food contact and chemically resistant applications. A further type of thermoplastic PU elastomer is used as a highly elastic fibre. This is known as spandex or elastane, or by its most common trade-name, Lycra.

4.1.2.2 Polycarbonate (PC)

One of the most popular materials for medical device manufacture is the engineering plastic polycarbonate. It has a range of properties which make it suitable to replace both glass (it is amorphous with glass-like transparency) and metal (strength, rigidity, and toughness) and it retains dimensional stability over a wide temperature range.

The clarity of PC allows it to be utilised where visibility is essential, such as when viewing blood, tissue or other bodily fluids. Some grades are also biocompatible, complying with ASTM and ISO biocompatibility testing standards, and are therefore suitable for filter cartridges in renal dialysis, blood filters used in heart bypass surgery, and connectors on fluid lines such as check valves and fittings.

A further advantage of PC is the ability to sterilise it by a number of methods, and its dimensional stability ensures leak-free operation.

The strength of PC is due to a large and bulky monomer structure incorporating aromatic rings, which limits molecular movement and imparts stiffness. Despite being amorphous it has comparable strength to crystalline engineering materials such as nylon and acetal. This strength is also due to forces of chain interattraction, as constituents of PC are chemically attracted to each other. This mix of clarity with inherent toughness is unrivalled amongst the thermoplastics and finds a variety of cross sector applications. On the downside PC has poor resistance to alkalis, and limited resistance to UV light.

4.1.2.3 Antimicrobial plastics

The use of antimicrobial plastics is growing rapidly. Here plastics are doped with additives, for example silver, which acts as a biocide. A biocide can be considered as a material that kills microbial organisms.

Other additives have a biostatic approach, which means that they prevent further reproduction of an organism, or some additives combine both biocide and biostatic actions. Arsenic-based materials are the most commonly used, although there are concerns over their toxicity. The uses of these additives spread across whole families of materials including PVC, PP, PET, ABS and PC and find applications in areas such as medical gloves. However, their use also extends beyond the medical sector into areas such as food preparation surfaces, floor tiles, toys and shower fittings.

4.1.3 Automotive applications

The amount of plastics incorporated in our cars has been steadily increasing, and while the amount varies from model to model this trend is likely to continue. Plastics currently make up about 12% of an average car.

The drive towards greater fuel efficiency, lower emissions, increased passenger safety and low cost has led to a wide variety of plastic materials and processes being used by automotive manufacturers. These range from super high strength composite materials replacing load bearing steel components to plastics used simply for aesthetic effect.

Interior trim, air bags, seat assemblies including foam padding, steering wheels, exterior body panels and bumpers, exterior trim and many under-bonnet components are all made of plastics.

Both thermoplastic and thermosetting materials and their composites are used. Materials include polyurethanes (PU) (thermoplastic and thermoset), thermoplastic elastomers, unfilled and filled (natural fibres, glass fibres, mineral filled) materials. Processing methods include thermoplastic processing techniques, GMT sheet moulding, injection compression moulding, and reinforced reaction injection moulding (RRIM) as well as assembly and finishing processes such as painting.

All common plastics processing routes described in Part 2 are used somewhere in the automotive sector. Advanced multi-material processes are also commonly used in this diverse sector, where low volume, custom cars, high quality luxury cars and high volume models all have different requirements.

The testing methods applicable in the automotive industry depend on where a certain material is used. For example an exterior body panel would need to be considered for weathering but an under-bonnet component would not. However, this part may need resistance to oils or heat etc. Often individual manufacturers develop their own specific material tests and standards for their materials.

It has been predicted that by 2020, a 50% reduction in automotive weight will have been achieved thanks to the demands for new fuel-efficient vehicles[b]. The weight reduction offered by replacing steel with plastics, along with the ability to consolidate a number of parts into one, continue to drive more plastic components in our vehicles.

Two areas of current development in this regard are automotive glazing and plastic body panels.

Automotive glass has long been a target for replacement with lighter weight material, and plastic glazing is now being introduced in concept vehicles by a number of manufacturers (Lincoln MKT, Land Rover LRX and Hyundai HED-5 i-mode). Polycarbonate is the material choice. Weight reductions of up to 50% can be achieved by replacing the current glass and metal systems, and previously unachievable structures can provide features such as panoramic roofs and wrap around designs. The window frames can be produced using PC/ABS blends.

Injection moulded body panels have already been used on a production vehicle. These were introduced on the Smart car as a changeable feature for the consumer. To obtain durability it is necessary to apply a thin layer of clear thermoset material to the thermoplastic panels. This is called a 'clearcoat' and aids scratch resistance.

Body panels big and small have now been made from a number of materials, including old PET bottles reinforced with glass fibre, PC/PBT blends, polypropylene and modified PPO. These have been introduced on a number of manufacturers' models including Land Rover, Citroen and Honda. Horizontal body panels compression moulded from 50% glass filled polypropylene composite have been used on the Hyundai QarmaQ concept vehicle, and these panels are 50% lighter than their steel counterparts. PP compounds have been widely used in the Honda Element model including front and rear fenders, side-fill garnishes, bumpers, and the tailgate. **Figure 4.3** shows a typical plastic car body panel incorporating a hole for the fuel inlet. Similar parts are made from plastic on a variety of models.

Removing the need to paint automotive components could reduce overall production costs by up to 25%, as well as this it would reduce carbon emissions, save energy and reduce the emission of volatile organic components (VOC). It is therefore not surprising that this is an area of great interest to manufacturers. There are a number of alternatives to painting. One way is to use high gloss, durable

b. US National Highway Traffic Safety Administration 'Enhancing future automotive safety with plastics', www-nrd.nhtsa.dot.gov/pdf/nrd-01/esv/esv20/07-0451-W.pdf, 2007.

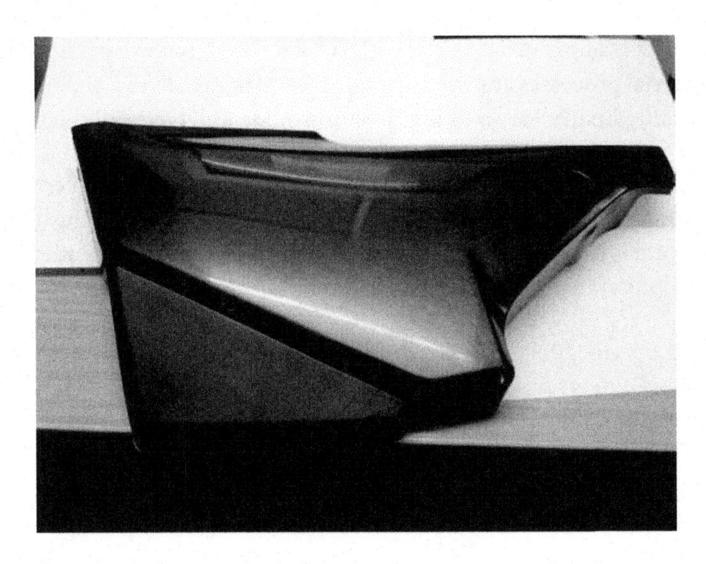

Figure 4.3

Automotive plastic panel incorporating hole for fuel flap/cover

plastic materials of the desired final colour, which give the high class and polished finish that is prized by many automotive companies. However the production of such glossy finished products without further finishing stages is technically very difficult. When injection moulding this requires the use of a highly polished mould tool cavity, and a suitable plastic capable of producing such as finish. For example glass filled materials tend not to be glossy. Other production processes such as extrusion and blow moulding do not tend to produce highly glossy surfaces on their components.

A further method of producing an unpainted surface is to use a laminate film of the desired decorative finish. The variations available here are endless and colours, metal finishes such as chrome effects, patterns and even photographs can all be reproduced. Again, the limitations are inherent in the finish in the film and the process it is applied to. When this technique is used with injection moulding, the film is placed in the mould tool cavity prior to each material injection. The process is often referred to as 'back injection' as the plastic is injected onto the back of the film. The technique can also be applied to surface injection mouldings, foils, textiles or wood. A wood laminate produced in this way and then lacquered to improve gloss is shown in **Figure 4.4**. This is a production component on a door interior in a Jaguar car.

Figure 4.4

Real wood effect injection moulding

4.1.3.1 Multi-layer fuel tanks

Fuel tanks have been made of plastic for a long time and compete with metal for this market. The ability of plastics to be moulded into a variety of shapes has allowed designers to fit these tanks into spaces that metal tanks would not have been able to occupy. These tanks can be produced by blow moulding in HDPE, but to ensure low fuel permeation, fluorinated HDPE is used. Fluorine replaces some of the hydrogen atoms on the surface of the polyethylene to improve barrier performance. Fluorination is also used to improve the barrier properties of monolayer bottles, but to obtain even lower permeation rates required for fuel tanks a multi-layer structure is used. By using six layers a further barrier layer (EVAL) can be incorporated into the structure as well as recycled material. An example of six layers is shown in **Figure 4.5**.

The adhesive layer (tie layer) is necessary to bond EVAL to HDPE; otherwise the layers would not stick together. It consists of a polyethylene material doped with a 'compatibiliser', typically maleic anhydride. Although all the layers are all the same thickness on the diagram, the tie layers would generally be thinner as these are expensive and need to be kept to a minimum.

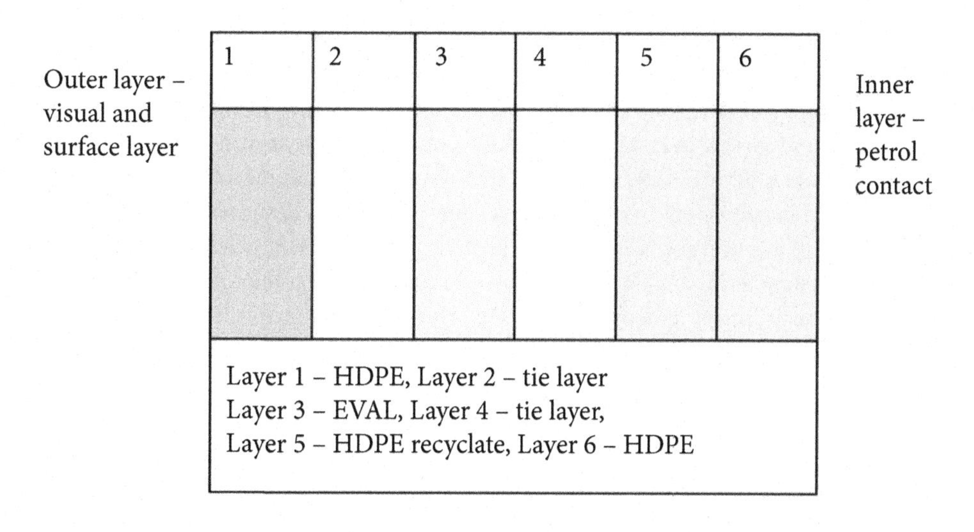

Figure 4.5

Multi-layer fuel tank structure

4.1.3.2 Composites and biocomposites

The automotive sector is one of the main business sectors for composite materials.

Replacing metals using composites is a continuing trend based on both weight reduction and cost reduction (or preferably both). One of the most recent examples on a production car is the running board on the General Motors Trailblazer. A five piece steel and plastic assembly has been replaced with a one-piece, long glass fibre-filled PP part, resulting in a 50% reduction in component weight.

A new material likely to find increasing use in the automotive industry has been developed by DuPont Engineering Polymers. Described as a nanocrystalline metal/polymer hybrid, the material called MetaFuse™ has the strength and stiffness of metal but allows designers all the freedom of a high performance thermoplastic. This is a nanotechnology whereby a thin layer of metal coats the components. This could target under-body engineering applications such as engine oil pans, transmission housings and steering components, replacing aluminium parts.

Bio-based materials are also starting to be incorporated into today's models and the automotive industry has helped champion these materials in its aim to meet stringent environmental legislation. Toyota has even built a pilot plant in Japan to manufacture its own biomaterial and aims to grow usage of this material to 20 million tons by 2020. Toyota also plans to have 15% of all its plastic parts made from either biomaterials or recycled materials by 2010. The advantages are claimed to be reduced oil use, reduced carbon emissions during use, reduced weight and improved recyclability. The Toyota Raum uses plastic made from sugar cane in the spare tyre cover and the floor mats.

Natural fibre usage in the automotive industry is also increasing. Again Toyota has been using kenaf fibre instead of glass fibre in some of its interior components.

This environmental leaning is however not by chance. Legislation has played a big part in the drive for car manufacturers to be seen as environmentally aware, and all manufacturers must comply with stringent environmental targets. The European Union Directive on end-of-life vehicles (ELV)[c] ensures that car manufacturers consider the fate of their vehicles once they have served their purpose. It has also put quotas on how manufacturers can dispose of their components, and forced car makers to design new models with the environment in mind. Old vehicles must go to designated dismantlers in order to disassemble the vehicle and recover high levels of various materials (such as metal, glass, plastic and oils). Any residual material must be disposed of in a controlled manner. As vehicle manufacturers have these stringent recycling and recovery targets to meet, this influences their material choices right from the design stages where this end of life 'design for disassembly' must be considered.

4.1.4 Electrical and electronic goods

Televisions, computers, mobile phones and fridges are just some of the everyday electrical and electronic items we use in our homes (**Figure 4.6**). The electronic and electrical sectors' total materials usage comprises about 30% plastics by weight and is rapidly growing.

In consumer electronics about twelve different plastics are routinely used, however the largest proportions comprise HIPS at 56%, ABS at 20% and poly(phenylene ether) (PPE) at 11%. In both TV sets and computers HIPS and ABS are routinely used for casings.

However given the growth in our use of electronics it is not surprising there is also concern at the growing volume of plastic and other waste in this sector. The lifetime of mobile phones is relatively

c. European Commission 2002. Directive 2000/53/EC on end-of-life vehicles (ELV). Official Journal of the European Union L269.

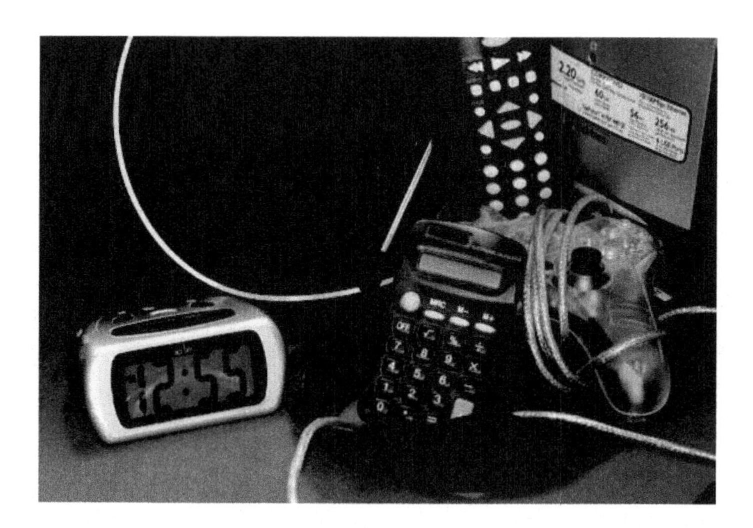

Figure 4.6

Examples of common electronic items containing substantial plastic components
Source N. Goodship

short for instance, and people also change or upgrade equipment such as computers far more quickly now than in the past. Therefore in response to a future waste problem, the European Commission adopted a directive addressing waste electrical and electronic equipment (WEEE)[d] defining recycling and recovery quotas for 10 WEEE categories, to cover metal and glass as well as plastic fractions.

4.1.4.1 High impact polystyrene (HIPS)

To remove the inherent brittleness of polystyrene and increase the impact performance, butadiene rubber can be introduced either by blending or during the polymerisation process (to produce a copolymer). The copolymerisation route produces improved properties over blending. This makes an inexpensive and lightweight polymer that can be used for numerous applications where high impact strength and good elongation are required. These materials typically contain 2-15% by weight of rubber component and consequently lose the clarity of natural polystyrene, having an opaque milky appearance.

HIPS can be used in all thermoplastic processing routes but is an ideal material for thermoforming. The high impact strength means it can be sawn and punched without difficulty.

In addition to electronic components, HIPS tends to find use in more demanding packaging applications, for medical packaging and also taste and odour sensitive products. It is also easily labelled and printed for products such as ID cards, name tags, dials, model making and signage.

d. European Commission, 2003. Directive 2002/96/EC on waste electrical and electronic equipment (WEEE). Official Journal of the European Union L37.

4.1.4.2 Acrylonitrile-butadiene-styrene (ABS)

This is a terpolymer made from three separate monomers which combine to produce a combination of high impact strength (butadiene rubber) and high mechanical strength (acrylonitrile and styrene). In addition to electronic components it is used for a range of applications such as automotive parts, pipes, telephone components, shower heads and door handles. For packaging it can be used for applications that need high tear strength such as lids and tubs. ABS can be readily electroplated and painted.

ABS plastics give better mechanical performance than the commodity plastics and also have good electrical insulation properties at a range of temperatures and frequencies.

4.1.4.3 Poly(phenylene ether) (PPE) – a heat resistant amorphous polymer

PPE is an example of a high end engineering polymer. PPE has very good mechanical properties (good impact strength and stiffness), good resistance to heat, excellent dimensional stability and very low creep behaviour at elevated temperatures. It has excellent electrical properties and when modified with flame retardant additives can make excellent housing materials for machines and appliances. PPE also has a low water absorption which makes it ideal for water contact applications.

On the downside PPE is not an easy material to process and needs high temperatures. It also has poor chemical resistance and poor colour stability. Post processing painting or electroplating is therefore needed to impart both added colour and chemical resistance to this material.

Typical applications may be internal appliance components (brackets and structural) such as the chassis of a telephone, large business machine (computer or printer) casings, high tolerance electrical switch boxes and connectors (including under-bonnet applications), industrial water pump connectors and pressure vessel housings.

Outside the electronic and electrical sector this material is also used for electroplated automotive wheel covers.

PPE is often used in a blend with either nylon or styrene to improve processing performance, whilst retaining dimensional stability, high heat performance and high mechanical strength.

4.1.4.4 Epoxy resins

As well as electronics applications, epoxy based materials find diverse applications in coatings, adhesives and composites. Epoxy materials can be modified to have a wide range of properties but are characterised by good mechanical properties, heat and chemical resistance. They are also known for their good adhesion characteristics.

Epoxies are naturally electrically insulating but like other polymers can be modified to have good electrical conductivity. For example by doping with silver both high electrical conductivity and strong conductive bonding can be achieved. These conductive epoxies can be used to replace soldering on heat sensitive materials. Epoxy adhesives of this type can have thermal conductivities of 1.8 W/mK.

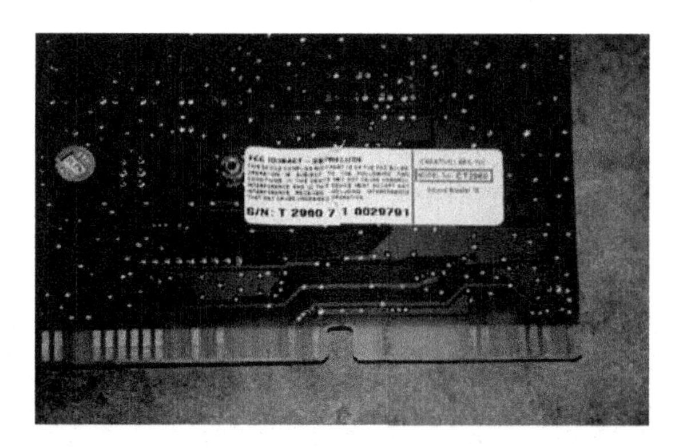

Figure 4.7

Epoxy circuit board

In the electronics industry, epoxies can be used as insulators to prevent short circuiting and they are used in circuit boards (as a composite sandwich with glass – see **Figure 4.7**) and transistors. Epoxies are also used to join and seat various electrical components within the boards.

4.1.5 Construction and structural engineering

In 2003, the construction market for plastics in the UK was estimated at about one million tonnes per year. The most popular material in this sector is PVC. (*Note the remarkable versatility of PVC, ranging from packaging to building to medical products!)* PVC is commonly used for many exterior applications such as window frames (**Figure 4.8**) and guttering, as well as interior flooring.

Other applications for plastics in construction include wiring, pipework, trimming, insulation, kitchen and bathroom components, cladding and conservatories. Polyethylene has good weatherability which makes it suitable for outdoor as well as indoor use in products such as drainage pipes (extrusion), signs, portable toilets (rotomoulding) and crates (injection moulding). However HDPE does undergo environmental stress cracking which is related to the crystallinity of polyethylene materials.

Figure 4.8

PVC window frames

Plastics offer designers the flexibility to innovate in future building design and construction, and this versatility (and the ability to make custom materials) are very attractive.

In civil and structural engineering, plastics and especially composite materials have become increasingly important materials. Construction is an important market for composite materials. For structural applications, thermoset reinforced materials are frequently used, and these are manufactured by processes such as filament winding, pultrusion and resin transfer moulding. In the US, maintenance of the transport infrastructure such as bridges is increasingly using composite materials. For extra stability in areas vulnerable to seismic activity, existing concrete structures are retrofitted using structural components made by filament winding to increase material ductility.

The first all-composite road bridge was built in 1996 in Kansas, using epoxy honeycomb structures. The bridge is 23 feet long and 27 feet wide. A number of other larger structures have since been build around the world as fibre reinforced polymers have become more commonplace in construction. Extruded honeycomb structures are often used in structural engineering, sandwiched between solid sheet materials (see **Figure 4.9**). The size and shape of the cells can vary but closely mimic the honeycomb structures produced by bees. They are made by extrusion followed by heat fusing of the blocks to produce a sheet. The uses of polypropylene honeycombs extend to hulls, bulkheads, door, wall and floor panels, wind energy blades, bridges, platforms and desks as well as reinforced packaging cases.

Reinforced composites in structural applications can withstand high loads, are fatigue resistant and corrosion resistant. The light weight aids the construction process making parts easy to manufacture, handle and assemble. Rot resistance may also be an issue in some construction components, and hence wooden components can be effectively replaced with composites for example.

The majority of fibre reinforced materials are generally manufactured with unsaturated polyester resins or epoxies and glass fibres. Reinforcing fibres such as carbon and aramid are also playing an increasingly important role.

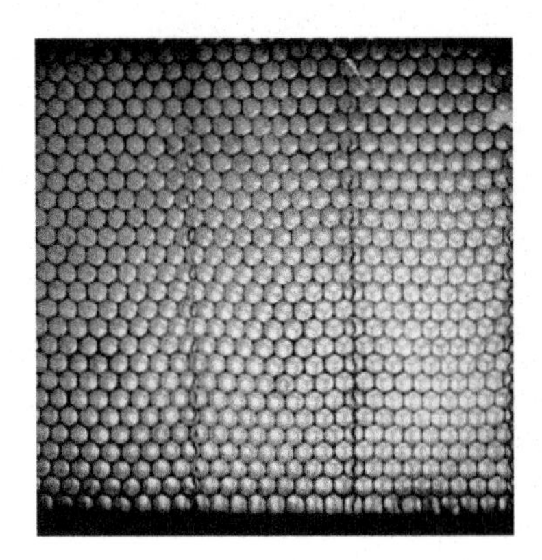

Figure 4.9

Honeycomb structure produced by bees (left), and made from polypropylene
Source N. Goodship

The aerospace industry is also a big user of composites, from aircraft wings to space shuttle components. In applications of this kind properties such as durability, weather resistance and difficulties in maintenance are all major factors in design choices.

4.1.5.1 Poly(vinyl chloride) (PVC)

PVC is a very versatile material. However, compared to a product such as a PVC bottle, a PVC window frame or pipe would be expected to have a much longer life cycle (decades rather than months). The durability requirements for these materials are therefore very different.

PVC has a similar monomer structure to both PP and PE, the difference being that PVC has a chlorine (Cl) atom attached to the polymer backbone instead of a hydrogen (H) or hydrocarbon (CH_3). The influence of packing and molecular structure on polyethylene was shown in Part 1, and the effect of the large chlorine atom is to reduce the ability for the chains to pack (like the effect of chain branching) making PVC amorphous rather than crystalline in character – the crystallinity levels are around 10%. A second implication of the presence of chlorine is due to the fact that chlorine is highly electronegative (which means it will attract electrons) and reduces the solvent resistance of PVC. A further difference from the polyolefins is an increase in tensile modulus and stiffness due to chemical attraction within the molecules. This also increases the glass transition (to around 60°C) compared to similarly arranged hydrocarbon materials. However, like other commodity plastics this is a low cost material and can be cheapened even more by adding fillers.

PVC can be processed using conventional thermoplastic methods but is heat sensitive. Therefore, specialised PVC machines are used commercially that keep the material at processing temperatures for a minimum time.

Rigid PVC, which is the unmodified version of this material, can be extruded to form window frames, gutters and conduits. It can also be thermoformed, injection moulded and blow moulded.

The plasticised version of PVC, which is also known as vinyl, is much more flexible. A plasticiser can be thought of as a partial solvent.

Imagine a glass of water in which sugar is dissolved. Now imagine a pile of sugar in which only a small amount of water is added *(it would probably look a bit of a sticky mess)*. In both of these cases the water is the solvent and the sugar is the solute. However in the second case only a small amount of solvent is added, so all the sugar does not dissolve. This is similar to what happens when PVC is plasticised. The polymer structure swells but does not dissolve. The individual chains therefore have more room in which to move and the flexibility of the polymer increases as does the softness. By varying the amount of plasticiser, a whole range of properties can be attained. This flexible PVC can be used for a whole range of applications such as plastic covers, flooring and electrical cables, as well as raincoats.

A further property common to both rigid and flexible PVC is flame retardance, and it will only burn reluctantly in the presence of a flame. If the flame is removed, PVC will cease to burn due to the chlorine content, a tendency which is termed self-extinguishing. This flame retardant property often leads to the specification of PVC in applications where other materials have seemingly superior mechanical properties. The flame retardancy can be further improved with the use of additives.

However, when PVC does burn, hydrochloric acid (HCl) is produced (as occurs if PVC degrades during processing). This does not preclude PVC from uses such as cable sheathing as it has been shown that the liberated hydrochloric acid breaks down very quickly in moisture, but HCl liberation from the uncontrolled mass incineration of PVC containing plastics would pose a greater health hazard.

Whether PVC use is a danger to humans has remained an item of political debate for several decades. For example the State of California (USA) proposed a bill banning its use as a packaging material, which was heard in 2008.

4.1.5.2 Expanded polystyrene (EPS)

This is polystyrene that has been foamed. Very highly foamed materials can contain as little as 2% PS, with the rest of the material being made up of air pockets. However 5% PS is more typical.

EPS has excellent thermal insulation properties and is therefore used to insulate buildings throughout the structure (roof, walls, and floors). It has high moisture resistance and can also be used to insulate at low temperatures for example in refrigeration or freezer facilities.

4.2 Green issues: reuse and disposal of plastics

It was shown in the first part of this book that the largest proportion of plastic is consumed by the packaging industry. For example, bottled water has become a worldwide phenomenon since its introduction in the mass market in the 1970s. The PET bottles that are used for packaging have therefore become more prevalent. The food and beverage market accounts for 60% of all packaging used and this market is expected to continue to grow in future. This 'white pollution' or 'white waste', of plastic bags, single use containers and other packaging has become a worldwide cause for concern. Some countries have banned the use of thin plastic bags; however over-packaging of products is seemingly ingrained in culture and economy. A retailer uses packaging as a sales tool for marketing and branding, to protect the contents (from damage and/or theft) and for delivery.

4.2.1 Reusing plastics

Reuse of products is one way of reducing the volume of the waste stream. For example, in the car industry the trade in second hand used parts has always been seen as a cost-effective practice for obtaining components. While this is generally not done with recycling in mind it is actually an example of reuse in action. A further example of reuse is when refillable containers are sold or when people refill old PET carbonated drinks and water bottles (despite the warnings on the bottle not to do so).

For gardeners, you can buy bottle tops which turn your old drinks bottles into plant watering cans (**Figure 4.10**). Again, reuse in action.

The widespread practice of landfilling our waste, as well as the emissions from incineration plants, have created many problems. Consequently highly visible plastic waste (especially packaging waste) has become a political issue. Environmental legislation has been introduced and now recycling is becoming commonplace and an accepted part of our society. All the industrial sectors discussed are now subject to plastics recycling legislation to some degree.

In 2001, the Alliance for Environmental Innovation in the US, produced some interesting data relating packaging energy consumption and material type, and this is shown in **Table 4.2**.

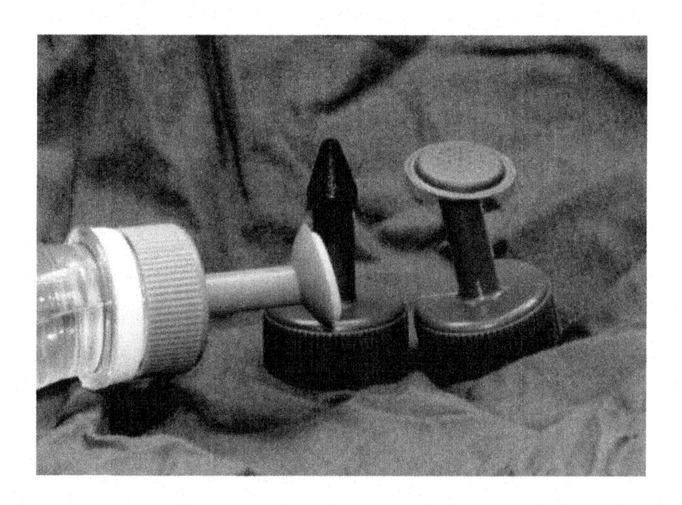

Figure 4.10

Reuse of bottles as plant watering cans

Source N. Goodship

Table 4.2

Energy requirements for manufacture of packaging materials

Material type	Energy Required (J/g)
Recycled glass	6,336
Recycled unbleached paperboard	7,392
Virgin glass	8,448
Recycled aluminium	10,560
Recycled HDPE	12,672
Virgin unbleached paperboard	23,231
Virgin HDPE	92,925
Virgin aluminium	192,186

What is notable about this chart is that the energy saving attached to the use of recycled aluminium can be clearly seen, and it is not surprising this material has such a well established recycled infrastructure. However, the material offering the second greatest energy saving is not paper or glass, also materials which are traditionally recycled. In fact it is HDPE, the material that we use to make our plastic milk bottles. The use of recycled HDPE can save us a significant amount of energy. A number of county councils in the United Kingdom have started to collect plastic bottles and a proportion of these currently collect only HDPE. This is one reason why.

However, there are other reasons for the difficulties faced in the recycling of plastics and their great variety and versatility is one of the main problems. There are so many different types of plastics both generically and individual grades of the same material (a grade is a specific mixture of additives, reinforcements and colours) that sorting of plastics is a major issue in their recycling.

However, before beginning to look at the issue of plastic waste it is worth considering ways that plastics can meet the first key step in waste management – source reduction. Reducing the amount of waste in the first place is the easiest way to address the problem. This can take the form of eliminating unnecessary materials, substituting more environmentally friendly ones, or reducing the weight of existing products.

Common products such as HDPE milk bottles and PET drinks bottles have seen regular weight reductions since they were introduced. Large HDPE milk containers are now over 40% lighter than they were when first introduced (95 g down to 60 g). Likewise, PET drinks bottles have also got steadily lighter, using less material and therefore generating less waste. Plastic bags have also become thinner. There are a number of drivers for this, not always environmental ones. For example, competitive pressures make savings in raw materials usage attractive. Also the materials themselves have advanced and become better performing. This allows less material to do the same job.

Substitution of a complex assembly with a single plastic component can simplify recycling, however sometimes plastics have replaced materials which are seemingly easier to recycle such as paper. Paper bags in supermarkets are much heavier than plastic bags, so there is a massive weight reduction in switching to plastic carrier bags. However, paper bags are easier to recycle as the paper recycling infrastructure is more widely developed than plastics recycling.

Recycled material based on waste discarded by the consumer is called post consumer recyclate (PCR). This is distinct from plastic manufacturing waste, which can generally be fed back into the manufacturing process. This process scrap remains relatively pure and easily identified in-house by the manufacture that produced it. Tonnes of plastic materials are routinely recycled in-house as general manufacturing practice. Excess manufacturing waste not reprocessed in-house is dealt with separately to PCR.

4.2.1.1 Sorting and mechanical recycling

As it is clearly feasible to recycle HDPE, you may be wondering by now why not simply collect and recycle all waste plastic? After all, aluminum cans are collected and recycled, as are paper, cardboard and glass – although glass is generally sorted by colour prior to collection. This last fact provides another clue to a problem with the recycling of plastics: their colour. Glass containers are generally (but not exclusively) just three colours: clear, green and brown. However plastic containers can be any of thousands of colours – another problem with versatility.

As an exception, think back to HDPE plastic milk containers, which are all the same colour – or more accurately colourless, the cloudy appearance is a property of natural HDPE. They are also a very common product, used every week by almost every household, and discarded immediately into the waste stream. They are easily recognised by the householder as being recyclable if a facility or collection exists, and easily distinguished by curbside recycling collectors such as provided by UK county councils. In this instance a relatively pure stream of recycled material is easily obtained.

The amount of material entering the waste stream, and the purity of it, are very important concepts in plastic recycling. This is because different types of plastics simply do not like each other. Their behaviour can be very similar to an oil and water mixture where the materials do not mix together. Therefore plastics are classed as being compatible (getting on) or incompatible (not getting on). The majority of plastics are incompatible with each other – although all the different types of polyethylenes are compatible. If incompatible plastics are mixed together the properties of the subsequent material are very poor.

A further problem comes with the different melting points of plastics. Imagine a waste stream containing HDPE, nylon 6 and PVC. A temperature required to reprocess HDPE, 200°C, would not melt a nylon material that was also in the waste stream. A temperature of 260°C, which would now melt the nylon as well, could cause PVC components in the waste stream to decompose releasing hydrochloric acid fumes. This higher temperature could also cause the HDPE to start to degrade, as 260°C is above the recommended processing temperature range for HDPE. Therefore we can see that it is necessary to sort the various plastics from each other, ideally by generic material type.

There are a number of methods of sorting waste plastics from each other. The easiest method is to do it at source (such as only taking HDPE bottles). However if mixed plastics enter recycling facilities, ways must be found to sort them in high volumes. This can be done manually, but is very labour intensive and operators can make mistakes, thus it is better to automate the sorting process.

One method is to use density separation. Remember that polyethylene and polypropylene float in water as they have a lower density. By using float and sink separators, the floating fraction of material (i.e. PE and PP) can be separated from the heavier materials such as PET and PVC.

PET and PVC have similar densities, but can be separated by a method called X-ray fluorescence (XRF), which uses differences in chemical composition to identify the materials. PVC contains chlorine which is detected by this method. Other methods routinely used to identify and separate various plastic fractions include infra-red absorption (IR) or methods which compare light transmission or absorption related to the inherent molecular structure of the polymer.

Once a waste stream is sorted into material types, these can be fed into suitable reprocessing streams. However there is always a proportion of the material that cannot be recycled this way, which will go on for further treatment methods.

To aid in manual sorting of plastics for recycling, components must be labelled with their generic code as shown in **Table 4.3**. Pick up any common plastic component and you should be able to find this mark.

Table 4.3
Recycling plastic numbering system

Recycling number	Plastic type	Abbreviation used on symbol
1 PETE	Poly(ethylene terephthalate)	PET (or PETE)
2 HDPE	High density polyethylene	HDPE
3 V	Poly(vinyl chloride)	PVC (or V)
4 LDPE	Low density polyethylene	LDPE
5 PP	Polypropylene	PP
6 PS	Polystyrene	PS
7 OTHER	Other	Other

Separated material can be reprocessed using many of the techniques described in Part 2 of this book once purity has been established. Individual processes have different tolerances to cope with recycled materials, however most can incorporate recycled materials in some form as long as the quality of the material (purity) is high. This can range from using 100% recycled material in a new product to incorporating just a small percentage of the total feedstock. The exception with respect to purity is intrusion moulding, which can cope with a highly mixed feedstock, but the resulting mouldings have poor properties.

The technologies described above all fall into the general category of mechanical recycling. The steps to mechanical recycling may include some or all of the following: collection, size reduction, washing (to remove residues), sorting, re-extrusion and re-processing, and each of these steps involves a cost. Mechanical recycling is not cheap and the product must then compete in the marketplace with clean virgin feedstocks.

4.2.1.2 Mechanical recycling in action

PET bottles are produced on a massive scale. In the United States alone 40 million plastic bottles are used annually. Processing of post consumer regrind (PCR) into higher value items is perhaps best demonstrated with the first revolutionary use of 90% recovered PET bottles. The clothing manufacturer, Patagonia, introduced their fleece made from these bottles in 1993, and recovered 86 million carbonated soft drink bottles from the waste stream in the next 13 years. It also uses PCR yarn in liner materials.

A further example of waste being used in a different manner to its first product life is seen in automotive body panels, again made from PET bottles. The PET recyclate was re-compounded with glass fibres for this application on the Smart car. PET bottles were also reused as part of the Hyundai QarmaQ concept car.

Same use applications are commonplace for many plastics. For example Innocent, the fruit smoothie manufacturer, uses recycled PET for their drink bottles. Used HDPE milk bottles can be fed back into the HDPE feedstock and used with virgin materials to create new milk bottles, and many plastic bags also incorporate a percentage of recyclate materials blended with virgin feedstock.

Other recycled materials can be used in the manufacture of water butts (see **Figure 4.11**), compost bins and plant pots. Black bin liners routinely incorporate recyclate materials. Being black is particularly helpful in allowing recyclates of different colours to be used and mixed, hidden by the black pigment.

Figure 4.11

Water butt made from recycled HDPE by rotational moulding

Source N. Goodship

However less than 20% of our plastics are currently mechanically recycled in

Europe, whilst the rest goes on to be thermally treated or landfilled. This includes many of the plastics designated 'Other', and also thermoset materials which cannot be melt processed a second time. The actual recycling and recovery rates of plastics in European countries vary widely. In 2007 Switzerland achieved virtually 100% recycling and recovery whereas Lithuania has rates of less than 10%.

4.2.1.3 Landfill and incineration with energy recovery

A landfill is where rubbish is simply buried. Despite increasing concern over the environmental effects of landfilling, this is still the most widely used method of waste disposal in Europe and the US. Landfill has the advantage of being very cheap and waste is hidden by being buried. However, uncontrolled landfills in the past have caused environmental problems. Methane gas is generated as waste decomposes. Lack of control over the liquid residues of decomposition called 'leachate' can contaminate groundwater supplies. There are also health issues associated with vermin and land subsidence.

Modern landfills must be carefully managed to ensure that gas generation is safely controlled and monitored. Landfills must also be lined with a suitable protective barrier to control run-off of leachates and protect water supplies. Despite their high visibility in waste going to landfill due to their low density, plastics are actually fairly inert materials. It is often the additives within them, rather than the polymers, that can cause difficulties if they come out of the polymer and into the leachate.

The major problem with landfills is they need space, and that space is dwindling as we continue to generate rubbish. In countries such as Japan where space is at a premium, landfilling has never been a prevalent form of waste disposal. Landfilling is also becoming more expensive as it becomes more regulated. One way of reducing the volume of rubbish going to landfill is to burn it first. This is done by incineration.

Incineration increases the bulk density of rubbish by combusting it to leave just ash residues. This ash must still be disposed of in landfills but it is now a more inert material, provided that the remaining levels of hazardous non-combustibles are within safe limits. In the past waste was just incinerated with little control on emissions from chimneys. However modern day incineration plants have complex cleaners and scrubbers to reduce any health damaging emissions. Furthermore, the combustible nature of much of the waste stream such as wood, paper and plastics (they are made from oil) means that massive amounts of energy can be generated during waste disposal.

This energy can be used to generate electricity and modern waste incineration plants can heat entire cities. Like landfills, there must be careful management to ensure all health hazards are kept to a minimum. This process reduces the volume of waste going to landfill as well as making viable electricity, and it is termed incineration with energy recovery. Denmark favours this method and disposes of about 80% of its plastic by energy recovery.

4.2.1.4 Thermal recycling methods

Rather than burning the entire waste stream, by selectively heating plastic components it is possible to generate useful chemical components which can be used to make new plastics. There are various ways to do this but the principle is generally termed feedstock recycling or chemical recycling. Despite pilot plants being commissioned, these techniques are not yet widely used as they tend to require very specific feedstocks. However, these areas are being actively researched. The way a material can be broken down is related to how it is polymerised in the first place.

When crude oil is refined various fractions are separated during a process called cracking. Depolymerisation processes work in similar ways to produce smaller molecules and monomers under controlled chemical conditions of heat and pressure. Hydrocarbons such as PE and PP can be broken down by this method as can materials such as nylon.

For the thermosets a process called pyrolysis can be used. This is thermal decomposition in an oxygen-free atmosphere. There are a variety of pyrolysis methods including fixed bed reactors, rotary kilns, screw pyrolysers and fluidised bed reactors.

4.2.2 Biodegradability

Recycling is one way to deal with plastic waste, another disposal route for some plastics depends on biodegradation in the environment. This could take the form of simple outdoor exposure to the elements whereby a product eventually falls apart, or placing it in a dedicated composting facility (or perhaps at some point in the future – simply burying it in your garden or adding it to your own compost heap.)

It has already been mentioned in Part 1 that this can be achieved with particular synthetic plastics such as PVAL or with degradation-inducing additives doped in conventional plastic materials such as PP and PE. However, to be compostable products must also fulfill specific requirements of designated standards such as ASTM D6400, ASTM D6868, EN 13432, ISO 14852, ISO 14855, and ISO 16929. These standards are more rigorous than those required to designate a product as biodegradable.

An example of a compostable material is a starch-based bioplastic material consisting of 100% renewable resources such as starch, fat and other additives. This can be decomposed by microbes into the soil within a composting facility in around 90 days leaving no harmful byproducts or contaminants. These materials are commercially available for applications such as packaging film, agricultural film, compostable bags and netting. Other possible applications (hygiene permitting) are surgical threads, gloves and aprons, nappies and similar products.

Compostable starch films are typically milky coloured and have low densities (around 1-1.2 g/cm^3). The starch is hygroscopic and needs to be kept dry. Processing temperatures are generally lower than those required for PE, but similar equipment can be utilised. Freshly produced materials are brittle but strength is regained after 24 hours, due to water content changes during processing. The major downside to starch-based materials is the cost. Competing materials such as PE and PP are among the cheapest of all plastics whereas fully compostable products tend to be very expensive.

Other starch based biodegradable materials are available for injection moulding of products such as cutlery, golf tees, plant markers, stakes and pegs, caps and closures, and extrusion of biodegradable pipes. Thermoforming can produce items such as plates, and disposable food trays (**Figure 4.12**). The mouldings always need to regain moisture content after processing for full strength to be achieved.

Some grades of these starch-based materials can also be used as doping agents to confer biodegradable properties on PE. For example commercial blends consisting of 55% starch, 20% PE and other additives can be used as an additive masterbatch into standard PE grades. Even if relatively high percentage levels of starch are retained in the final component (e.g. as much as 60%) there is still a considerable price reduction compared with using pure compostable materials, as cheaper PE makes up a percentage of the final mixture. However compared to PE alone, these resin blends are still very expensive but they have the benefit of biodegradability. Full properties are not obtained until the starch component has regained moisture after processing, so films need to be left 24 hours before use. The blend materials do not retain the full properties of PE, as there is a reduction in elongation and mechanical performance, however this change can be tolerated in some non-critical applications.

The most widely used bioplastic is poly(lactic acid) (PLA), see **Figure 4.13**. This is made from corn starch or sugar cane and it is a biodegradable product with well known properties. It will degrade in composting facilities and is available in grades for extrusion (sheet and films) and injection moulding.

PLA must be pre-dried prior to processing as it is water sensitive and care must be taken not to contaminate the material with other plastics such as polyolefins. Processing temperatures are similar to those used for polyolefins, although generally temperatures above 200°C are not tolerated for very long.

PLA is transparent and it can be used for dairy and other food containers, plates, cold drink cups (See **Figure 4.14**), handles and screw caps. Manufacturing scrap can be reground and recycled back into production as long as it is dry. A comparison of its properties with other packaging materials is given in **Table 4.4**. It can be seen that PLA has strength properties comparable to those of PS and may compete with PET in some applications.

The PLA cup in **Figure 4.14** was used at the Wychwood music festival in the UK in 2008. It was printed with the words 'This is not a plastic cup'. However, it was still deposited into the plastic recycling bins

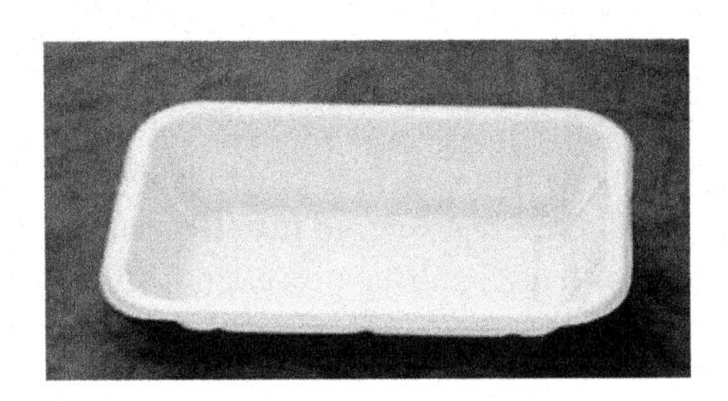

Figure 4.12

Starch based disposable food tray

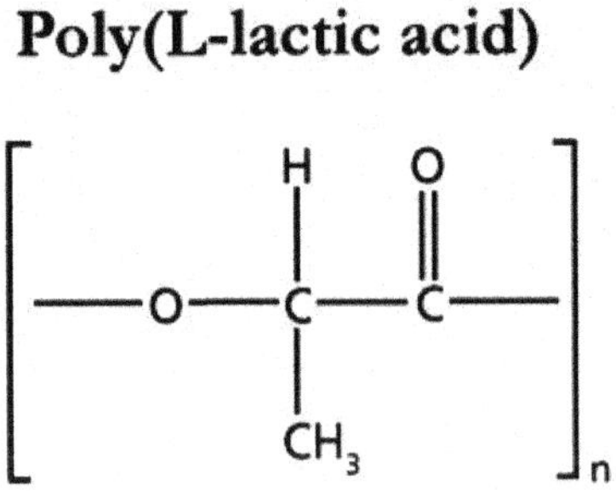

Poly(L-lactic acid)

$$\left[-O-\underset{CH_3}{\overset{H}{C}}-\overset{O}{\overset{\|}{C}}- \right]_n$$

Figure 4.13

PLA repeat unit

Figure 4.14

PLA drink cup, marked 'This is not a plastic cup'

Table 4.4
Indicative values of PLA properties versus other packaging materials

Property	Unit	PLA	PET	LDPE	PS
Density	g/cm^3	1.25	1.37	0.92	1.05
Melting point	°C	145-155	267	105-110	240
Glass transition temperature	°C	60	69-77	-125	110
Tensile strength	MPa	45	54	17-26	50
Elongation	%	3	50-300	>100	3-4
Impact strength	KJ/m^2	1-3	No break	No break	5-20

on site by consumers keen to support recycling. This demonstrates the educational and infrastructure confusion surrounding plastic recycling versus degradable and/or compostable plastics. In the future the presence of PLA in the plastics market could potentially encourage well-meaning consumers to throw all plastics away, thinking they are degradable. The actual presence of this material in the market at the time of writing is relatively small but a change in consumer thinking could actually damage the existing recycling infrastructure. Equally, if PLA contaminates the recycling waste streams it devalues the other plastics within it as it degrades their properties. As commercial composting facilities for materials such as PLA are not yet widely available it is likely that materials such as PLA will end up in landfill sites or recovered for energy until suitable infrastructure is put in place. Some of the issues around the use of PLA are summarised in **Table 4.5**.

Ethical issues still surround the use of biomaterials. They still use energy in their manufacture just as synthetic materials do. They also potentially divert food reserves from those less able to feed themselves. However, short-term issues such as these are likely to be solved for future generations of biomaterials. It is hard to argue that biomaterials such as these will not be part of our future in the longer term as we harness nature and advance our understanding of materials.

Table 4.5
Summary of advantages and disadvantages of using PLA

• PLA plastic is biodegradable and can be composted (and recycled if not biodegraded) • It is made from renewable sources i.e. corn starch or sugar cane • PLA can substitute for some synthetic polymers and is suitable for food packaging • It can be used for disposable plates and cutlery • It is appealing for eco-consumers who prefer to buy low environmental impact products
• PLA bottles in waste recovery streams could disrupt the existing successful recycling system by contaminating recycled feedstock • The degradable nature of PLA could lead to more consumers throwing away plastics rather than recycling because they believe plastics are all easily degradable • PLA plastic uses energy in its production and then turns back to biomass • Polymer production could potentially divert land and reserves from food production

4.3 Nature's polymer processing

This book has covered polymer and plastic materials, their testing and applications within the scope of a man-made world of high temperatures and high pressures. However as mentioned at the start, the natural world has been processing polymers such as DNA, proteins and cellulose without the need for such extreme processing environments. So while material scientists have come a long way, there is still plenty of room for improvement.

Consider the spider – a master polymer processor.

Spider web threads are stronger in proportion to their thickness than Kevlar, and they are spun in their own natural version of polymer fibre spinning. Spinning polypropylene requires temperatures of around 200°C, as well as high pressures and expensive manufacturing equipment. The humble spider spins a far superior product to polypropylene, and also aligns it to construct a web. Not only does the spider spin this fibre, but actually different 'grades' of the fibre depending on the requirements. The spider controls crystallinity and molecular alignment to change properties as it spins. Just to add insult to injury, there are no high temperatures required, and no expensive processing equipment!

The spider is not the only example of nature's control over processing polymer materials, but is perhaps the best known.

Will our polymer materials of the future be processed at room temperatures and pressures and from nature's own renewable raw materials?

Will they be built from a 'molecule up' approach as we increase our understanding of nanotechnology?

Will we combine polymeric and ceramic materials to create super strong and flexible materials as nature does?

The possibilities for new naturally made materials are very exciting and the technology is certainly within our grasp. Future generations of material scientists may one day laugh at our primitive attempts at plastics processing described in this book – I certainly hope so!

Appendices

Glossary of abbreviations used in this book

(see Tables 1.18 and 1.19 for abbreviations used for common polymers and copolymers)

ASTM	American Society for Testing and Materials, now called ASTM International
BOPP	Biaxially orientated polypropylene (film)
BS	British standard
CLTE	Coefficient of linear thermal expansion
DIN	German Institute for Standardization (Deutsches Institut für Normung)
DSC	Differential scanning calorimetry
ELV	End of life vehicle
GF	Glass fibre
GMT	Glass mat reinforced thermoplastic
GRP	Glass reinforced plastic
HAZ	Heat affected zone
HDT	Heat distortion temperature
IR	Infrared
ISO	International Organization for Standardization
JIS	Japanese Industrial Standards
L/D	Length to diameter ratio
LTM	Liquid transfer moulding
MI	Melt index
MFI	Melt flow index
MFR	Melt flow rate
NAFTA	North American Free Trade Agreement
PCR	Post consumer regrind
RIM	Reaction injection moulding
RRIM	Reinforced reaction injection moulding

RTM	Resin transfer moulding
SRIM	Structural reaction injection moulding
T_g	Glass transition temperature
TGA	Thermogravimetric analysis
T_m	Melting temperature
TMA	Thermomechanical analysis
TPO	Thermoplastic olefin
TPV	Thermoplastic vulcanate
UL	Underwriters Laboratories (product safety certification organisation)
UV	Ultraviolet
VOC	Volatile organic content
VST	Vicat softening test
WEEE	Waste electrical and electronic equipment
XRF	X-ray fluorescence

Some useful units and derivations

Property	**Unit**
Electrical Resistance	Ohm (Ω) , *this is the Greek symbol omega*
Electrical conductivity	$(\Omega\ m)^{-1}$
Energy	Joule (J), 1 J = 1 Newton metre (Nm) =1 kg $m^2\ s^{-2}$
Force	Newton (N) = kg m s^{-2}
Mass	Kilogram (kg)
Time	Second (s)
Stress	Frequently quoted in Pascals (Pa) 1Pa=N m^{-2} 1 Million Pascals (1 MPa) =1 MN m^{-2} = 1N mm^{-2}
Volume	Litre (L or l) 1 L = 1 dm^3 = $10^{-3}\ m^3$
Temperature	Kelvin (K) the SI unit of temperature (To convert to Celsius (°C) = K –273.15)

Glossary of common plastics-related terminology used in this book

Ageing	The effect of an environment (natural or artificial) on plastic over time
Amorphous	Random, disordered structure in a polymer
Axes of rotation	Rotational moulding term for various directions of mould tool movement
Barrier screw	A type of extruder screw configuration which improves mixing
Biopolymer	Naturally occurring polymer
Blow pin	Used in blow moulding to put inflating air into the moulding
Brittle	Having glass-like properties on breaking
Chalking	Powdery chalk-like deposit on the surface of plastics
Clamping force	The amount of force available to hold a mould tool shut
Clamping unit	The part of a moulding machine which houses the clamping mechanisms
Coefficient of expansion	Change in size over time
Composite	Physical mixture of polymer (matrix) and reinforcement (dispersed phase)
Compressive strength	A measure of a material's ability to resist crushing or compressive forces
Conductive polymer	Polymer with (electrical/magnetic) properties similar to metals
Copolymer	Polymer made up of two or more different monomers
Crazing	Series of small cracks on the surface or through a material
Creep	Material deformation over time while under load
Crosslinking	Chemical bonding between polymer chains
Crystalline	Ordered structure in a polymer
Die	The shaping part of an extruder where the material exits
Draw ratio	The amount a material can be stretched without splitting during thermoforming
Ductility	The amount a material can be plastically deformed without breaking
Dynamic load	A cyclic or intermittent load
Elasticity	The ability to recover original shape once deformation is removed
Elongation	Increase in length relative to original length, expressed as a percentage
Fatigue endurance	Ability to resist repetitive short-time stress, e.g. from cyclic loading
Fixed platen	The part on which the stationary part of the mould tool is seated, on injection and compression machines
Flexural modulus	The ratio of stress over strain of a material under flexural deformation
Flexural strength	A measure of the ability to resist bending under load
Gloss	The shine from a surface

Godet	Take-up rollers in melt spinning operations
Hardness	Resistance to penetration by indentation
Haze	Cloudiness of a transparent material
Heat affected zone	Welding term for the area affected by the weld
Homopolymer	Polymer made of just one monomer type
Hygroscopic	The tendency of a material to absorb moisture from the air
Impact resistance	The ability of a material to absorb energy resulting from dropping or striking
Injection moulding cycle	The process of producing one complete part from start to finish, and the time this takes
Injection unit	The part of a moulding machine that houses the injection screw and feed unit
L/D ratio	Ratio of length to diameter in extruder screws
Living hinge	A thin flexible hinge integrally moulded between two parts of a component
Macromolecule	A large molecule
Melt flow index (MFI)	A measure of material flow
Monomer	A single molecule – the building block of polymers
Micro moulding	Injection moulding of precision components with weights in the milligramme range
Moving platen	The part on which the moving part of the mould tool is seated, on injection and compression machines
Nip region	Calendering term for the region in which sheets are fed between rollers to compress and stretch them
Oxidation	Type of chemical reaction where there is a gain in the amount of oxygen at a molecular level
Parison	Blow moulding term for the extruded component before inflation
Passive screw	A transfer screw that serves no major mixing function
Photooxidation	Oxidation initiated by light
Plastic	Processable material comprising polymer and additives
Plasticity	Ability to deform without breaking
Polymer	Structure made up of chains of monomers
Preform	A shaped plastic structure not in its final form
Rheology	The study of flow under heat and pressure
Rigidity	Resistance to bending
Shot weight	The controlled dose of material required for one moulding
Spinneret	Name of the die used in melt spinning to make separate fibres
Static load	A non-varying load

Stiffness	Resistance to deformation
Stress	The load on a material
Strain	Deformation caused by stress
Stress cracking	Small cracks in a material caused by stress
Synthetic polymer	Man-made polymer
Tensile modulus	The ratio of stress over strain of a material under tensile deformation
Tensile strength	A measure of the ability of a material to resist a pulling force
Tie bars	Structural bracing parts of the clamping unit, on injection or compression moulding machines
Viscosity	Resistance to flow
Yield	The mechanical transition from elastic to plastic deformation under stress

Symbols and prefixes for frequently used multiples

Symbol	Prefix and multiplication factor
n	Nano 10^{-9}
µ	Micro 10^{-6}
m	Milli 10^{-3}
k	Kilo 10^{3} (this is used in lower case e.g. kg not Kg)
M	Mega 10^{6} (e.g. as used in MPa)
G	Giga 10^{9} (e.g. as used in GPa)

Commonly quoted standards related to plastics

ASTM D1238 - 04c	Standard Test Method for Melt Flow Rates of Thermoplastics by Extrusion Plastometer
ASTM D1598 - 02	Standard Test Method for Time-to-Failure of Plastic Pipe Under Constant Internal Pressure
ASTM D1746 - 09	Standard Test Method for Transparency of Plastic Sheeting
ASTM D256 - 06ae1	Standard Test Methods for Determining the Izod Pendulum Impact Resistance of Plastics
ASTM D3763 - 08	Standard Test Method for High Speed Puncture Properties of Plastics Using Load and Displacement Sensors
ASTM D3835 – 08	Standard Test Method for Determination of Properties of Polymeric Materials by Means of a Capillary Rheometer
ASTM D5420 - 04	Standard Test Method for Impact Resistance of Flat, Rigid Plastic Specimen by Means of a Striker Impacted by a Falling Weight (Gardner Impact)
ASTM D638 - 08	Standard Test Method for Tensile Properties of Plastics
ASTM D6400 – 04	Standard Specification for Compostable Plastics
ASTM D6868 – 03	Standard Specification for Biodegradable Plastics Used as Coatings on Paper and Other Compostable Substrates
ASTM D790 – 07e1	Standard Test Methods for Flexural Properties of Unreinforced and Reinforced Plastics and Electrical Insulating Materials
BS 6233:1982 (IEC 60093:1980)	Methods of test for volume resistivity and surface resistivity of solid electrical insulating materials
BS EN 60112:2003 (IEC 60112:2003)	Method for the determination of the proof and the comparative tracking indices of solid insulating materials
BS EN 60243-1:1998 (IEC 60243-1: 1998)	Electrical strength of insulating materials –Test methods – Part 1: Tests at power frequencies
EN 13432: 2000	Packaging – requirements for packaging recoverable through composting and biodegradation – Tests scheme and evaluation criteria for the final acceptance of the packaging
ISO 1133:2005	Plastics – Determination of the melt mass-flow rate (MFR) and the melt volume-flow rate (MVR) of thermoplastics
ISO 11357-1:1997	Plastics – Differential scanning calorimetry (DSC) – Part 1: General principles

ISO 11358:1997	Plastics – Thermogravimetry (TG) of polymers – General principles
ISO 11359-2:1999	Plastics – Thermomechanical analysis (TMA) – Part 2: Determination of coefficient of linear thermal expansion and glass transition temperature
ISO 1183-1:2004	Plastics – Methods for determining the density of non-cellular plastics – Part 1: Immersion method, liquid pyknometer method and titration method
ISO 14852:1999	Determination of the ultimate aerobic biodegradability of plastic materials in an aqueous medium – Method by analysis of evolved carbon dioxide
ISO 14855-1:2005	Determination of the ultimate aerobic biodegradability of plastic materials under controlled composting conditions – Method by analysis of evolved carbon dioxide – Part 1: General method
ISO 16929:2002	Plastics – Determination of the degree of disintegration of plastic materials under defined composting conditions in a pilot-scale test
ISO 178:2001	Plastics – Determination of flexural properties
ISO 179-2:1997	Plastics – Determination of Charpy impact properties – Part 2: instrumented impact test
ISO 180:2000	Plastics – Determination of Izod impact strength
ISO 22007-1: 2009	Plastics – Determination of thermal conductivity and thermal diffusivity – Part 1: General principles
ISO 2577:2007	Plastics – Thermosetting moulding materials – Determination of shrinkage
ISO 294-1:1996	Plastics – Injection moulding of test specimens of thermoplastic materials – Part 1: General principles, and moulding of multipurpose and bar test specimens
ISO 294-4:2001	Plastics – Injection moulding of test specimens of thermoplastic materials – Part 4: Determination of moulding shrinkage
ISO 306:2004	Plastics – Thermoplastic materials – Determination of Vicat softening temperature (VST)
ISO 307:2007	Plastics – Polyamides – Determination of viscosity number
ISO 4582:2007	Plastics – Determination of changes in colour and variations in properties after exposure to daylight under glass, natural weathering or laboratory light sources

ISO 4892-1:1999	Plastics – Methods of exposure to laboratory light sources – Part 1: General guidance
ISO 527-2:1993	Plastics – Determination of tensile properties – Part 2: Test conditions for moulding and extrusion plastics
ISO 604:2002	Plastics – Determination of compressive properties
ISO 62:2008	Plastics – Determination of water absorption
ISO 6603-2:2000	Plastics – Determination of puncture impact behaviour of rigid plastics – Part 2: Instrumented impact testing
ISO 75-1:2004	Plastics – Determination of temperature of deflection under load – Part 1: General test method
ISO 868:2003	Plastics and ebonite – Determination of indentation hardness by means of a durometer (Shore hardness)
ISO 877-1:2009	Plastics – Methods of exposure to solar radiation – Part 1: General guidance
ISO 899-1:2003	Plastics – Determination of creep behaviour – Part 1: Tensile creep
UL 94	UL Standard for Tests for Flammability of Plastic Materials for Parts in Devices and Appliances

Sources of additional information

Polymers and plastics

Grosberg A.Y. and Khoklov A.R., Giant Molecules: Here, There and Everywhere…, Academic Press, 1997

Brydson J.A., Plastics Materials, 7th edition, Butterworth-Heinemann, 1999

Elias H.G., An Introduction to Plastics, 2nd edition, Wiley, 2003

Sperling L.H., Introduction to Physical Polymer Science, 4th edition, Wiley, 2006

Woodward A.E., Understanding Polymer Morphology, Hanser Gardner Publications, 1995

Processing

Chandra M. and Roy S.K., Plastics Technology Handbook, 4th edition, CRC Press, 2007

Crawford R.J., Plastics Engineering, 3rd edition, Butterworth-Heinemann, 1998

Wilkinson A.N. and Ryan A.J., Polymer Processing and Structure Development, Kluwer Academic Publishers, 1999

Goodship V., ARBURG practical guide to Injection Moulding, Rapra Technology, 2004

Osswald T., Turng L. and Gramann P., Injection Molding Handbook, 2nd edition, Hanser, 2007

Rauwendaal C., Polymer Extrusion, 4th edition, Hanser, 2001

Henson F., Plastics Extrusion Technology, Hanser, 1997

Michaeli W., Extrusion Dies for Plastics and Rubber, 3rd edition, Hanser, 2003

Cantor K., Blown Film Extrusion: an Introduction, Hanser, 2006

Lee N.C., Practical Guide to Blow Moulding, Rapra Technology, 2006

Belcher S. L., Practical Guide to Injection Blow Molding, CRC Press, 2007

Throne J., Understanding Thermoforming, 2nd edition, Hanser, 2008

Beall G. and Throne J., Hollow Plastic Parts, Hanser, 2004

Crawford R.J. and Kearns M.P, Practical Guide to Rotational Moulding, Rapra Technology, 2003

Starr T.F., Pultrusion for Engineers, Woodhead Publishing and CRC Press, 2000

Yu S.Y., Mai Y.W. and Lauke B., Science and Engineering of Short Fibre Reinforced Polymer Composites, CRC Press, 2009

Troughton M.J., Handbook of Plastics Joining - a practical guide, William Andrew, 2008

Shah V., Handbook of Plastics Testing and Failure Analysis, 3rd edition, Wiley, 2007

Additives

Murphy J., Additives for Plastics, 2nd edition, Elsevier Science, 2001

Schiller M., Maier R. and Zweifel H., Plastics Additives Handbook, Hanser, 2009

Charvat R., Coloring of Plastics, Fundamentals, Wiley, 2003

Design

Lefteri C., The Plastics Handbook, Rotovision, Hove, UK, 2008

Applications of plastics

Selke S.E.M., Understanding Plastics Packaging Technology, Hanser Gardner Publications, 1997

Hernandez R.J., Culter J.D. and Selke S.E.M., Plastics Packaging: Properties, Processing, Applications and Regulations, 2nd edition, Hanser, 2004

Stauber R. and Vollrath L., Plastics in Automotive Engineering, Hanser, 2007

Akovali G., Polymers in Construction, Rapra Technology, 2005

Recycling and environment

Andrady A., Plastics and the Environment, Wiley, 2003

Azapagic A., Emsley A. and Hamerton I., Polymers, the Environment and Sustainable Development, Wiley, 2003

Bastioli C., Handbook of Biodegradable Polymers, Rapra Technology, 2005

Scott G., Degradable Polymers: Principles and Applications, Springer, 2nd edition, 2003

Rudnik E., Compostable Polymer Materials, Elsevier, 2007

Imhoff D., Paper or Plastic: Searching for Solutions to an Overpackaged World, Sierra Club Books, 2005

Goodship V., Introduction to Plastics Recycling, 2nd edition, Smithers Rapra, 2007

Scheirs J. and Kaminsky W., Feedstock Recycling and Pyrolysis of Waste Plastics, Wiley, 2006

Useful websites

The links from these sites often also link to other useful web resources

British Plastics Federation (BPF): www.bpf.co.uk

Association of Plastic Manufacturers Europe: www.plasticseurope.org

Society of Plastics Engineers (SPE): www.4spe.org

Macrogalleria: http://pslc.ws/mactest/index.htm

Wasteonline: www.wasteonline.org.uk

Americanchemistry plastics division: www.americanchemistry.com/Plastics/

Index

I

J

K

L

M

N

O

P

Q

R

S

T

Printed in the United States
By Bookmasters